Praise for

'*Show Me the Bodies* is a st⟨
to the victims, the bereave⟨
painstaking, forensic investi⟨
Yet it is also an unflinching portrait of Unp⟨⟨⟨⟨
privatized and neglected kingdom where profit for the few always
triumphs over the health, safety and lives of the many, where the
victims are always left voiceless, and where the dead never find justice
or peace. And where, most damningly of all, we still choose not to act
and so still let crimes such as Grenfell happen, over and over, again
and again. In short, this is the most harrowing, moving, powerful and
important book of the year, and one which every citizen should read.
And remember. And learn from and then act upon.'

David Peace, author of the *Red Riding Quartet*

'Working from painstaking daily reporting from the inquiry, alongside extensive interviews with the bereaved and survivors of the
Grenfell atrocity, Apps has written a concise, devastatingly detailed
and upsetting book ... From architects to politicians, all decision
makers should read *Show Me the Bodies*.'

Emma Dent Coad, former MP for Kensington

'Peter Apps has written a searing indictment of what he rightly calls
"the most serious crime committed on British soil this century"
in this forensic account of the deregulation, cost-cutting and sheer
negligence behind the Grenfell fire and its human cost. It's essential
reading if we are to avoid such needless tragedy in the future.'

John Boughton, author of *Municipal Dreams*

'Compelling, rigorous, utterly forensic and so very needed. This
book has to be the moment that things change.'

Lucy Easthope, author of *When the Dust Settles*

SHOW ME
THE BODIES

How We Let
Grenfell Happen

Peter Apps

ONEWORLD

A Oneworld Book

First published by Oneworld Publications Ltd in 2022
Reprinted in 2022, 2023 and 2024

ISBN 978-0-86154-615-2
eISBN 978-0-86154-595-7

Typeset by Geethik Technologies
Printed and bound in Great Britain by Clays Ltd, Elcograf S.p.A.

Oneworld Publications Ltd
10 Bloomsbury Street
London WC1B 3SR
England

MIX
Paper | Supporting
responsible forestry
FSC
www.fsc.org
FSC® C018072

For the 72 and all who love them

CONTENTS

INTRODUCTION

It should have been a normal flat fire. It was just an electrical appliance malfunctioning in a flat on the sixth floor of a 1950s council block. The London Fire Brigade (LFB) attend these sort of events every day. Often, they put it out without the other residents of the building even knowing. But this fire would be different.

The tower block had been poorly maintained and serious fire safety defects had been allowed to fester. Residents had raised their concerns without any success. A legally required risk assessment had not been carried out. Worse, a recent refurbishment had seen highly combustible panels fixed to the external wall.

It was the middle of a hot summer when the fire broke out, the flames licking through an open window, igniting one of the panels. It began to spread up the building, threatening other flats.

This took the fire service by surprise. Fire is not supposed to spread from flat to flat. As call after call came in from trapped residents, the call handlers fell back on the textbook advice: 'stay put'. On the ground, the rescue operation became chaotic. This was outside the firefighters' training and they didn't know how to respond. Outdated equipment hindered the co-ordination of the response. Command was passed rapidly from one officer to another. Key information necessary to save the trapped residents was not conveyed to the teams on the ground quickly enough. Residents were left waiting desperately for help that never came. If they had been told to flee, they would likely have lived.

Harrowing 999 calls, which would later be played at a mammoth public inquest, recorded the rising panic of those trapped as smoke filled their burning flats. The fire ripped through the poorly maintained building. Fire doors failed. Eventually, the single staircase filled up with pitch-dark, choking smoke. In just one bathroom, two mothers and their three children died, including a baby born just weeks before.

'The council were aware of our concerns. We told them we needed certain measures put into place,' one resident told the *Evening Standard* just days after the fire. 'But every time we complained, they told us they had taken our concerns on board, but nothing ever happened.'[1]

Questions rapidly emerged about other social housing tower blocks around the country, as it appeared some of the safety issues which had turned this fire into a disaster were widespread. Amid a storm of criticism, the fire service said it would review the stay put advice it had given trapped residents. It was Britain's worst ever tower block fire. Politicians solemnly promised it would never happen again. These promises would be broken.

Because this wasn't the Grenfell Tower fire of 2017. It was a fire at Lakanal House, a tower block in Southwark in south east London, in 2009.

Three adults, two children and one twenty-day-old baby lost their lives in a terrifying, painful, public and avoidable tragedy. But the bodies of Helen Udoaka and her daughter Michelle, Dayana Franciquini and her children Thais and Felipe and Catherine Hickman were not enough to persuade this country to change course.

The fire that killed them demonstrated all the fundamental flaws that would lead to the horrors seen at Grenfell Tower eight years later: combustible external panels allowed the blaze to spread out of the flat of origin and climb up the building's exterior; internal breaches to fire protection allowed smoke and flame to spread through the building; and the fire brigade's blind faith in its advice to stay put resulted in the victims losing their chance to escape.

All of this was known well before flames engulfed Grenfell Tower. We simply chose not to act. Regulations went unamended and plans to avoid a repeat of the disaster were not made. Warning voices were ignored.

It tells us something about how we are governed, and the priority our political and economic system places on human life – especially when those lives are likely to be poor, immigrant and from ethnic minority backgrounds. Grenfell was the result of a series of choices, the sum of state neglect and corporate wrongdoing across a variety of areas, the epicentre of myriad defects in our social fabric. More than anything, it was a result of political choices. 'Grenfell is a lens through which to see how we are governed,' said Stephanie Barwise QC, a lawyer representing the bereaved and survivors of the Grenfell Tower fire in December 2021.[2]

This book will peer through this lens. The picture it unveils should appal us. Over a period of at least thirty years, our representatives chose time and again not to act on mounting evidence that something needed to be done to prevent a disaster in a high-rise building. They deliberately ran down, neglected and privatised the arms of the state that might have otherwise avoided the need for this book. And they allied themselves with a corporate world that evinced an almost psychopathic disregard for human life.

If you think this is hyperbole, consider the below – all of which will be explained in the pages to come. An employee at one insulation manufacturer described a fire test on a system containing its product as 'a raging inferno' in 2007, but buried the testing and marketed it specifically for use on high-rise buildings. When its fire performance was questioned, one manager said those raising issues could 'go fuck themselves'. Senior figures at a cladding manufacturer exchanged emails internally saying the company was 'in the "know"' about its product's poor performance, but told its salespeople to keep its true fire performance 'VERY CONFIDENTIAL!!!!!'. And one internal document speculates about the commercial consequences of a fire in a tower block clad with its product killing '60/70 persons'. Both those materials would end up on Grenfell Tower.

Consider a contractor whose staff bragged about being 'quids in' due to the additional profit margin they would gain by switching the cladding on the tower for the cheapest option. Consider another contractor dismissing the need for tough fire breaks by writing 'as we all know; the ACM [cladding] will be gone rather quickly in a fire.' Consider the management company that concealed reports of a faulty smoke extraction system and wrote 'let us hope our luck holds and there isn't a fire.' Consider the fire service that fretted about issuing a public warning about the widespread use of combustible cladding in the capital as that would 'let the cat out of the bag'.

Then consider a government whose officials internally laughed off calls for tighter fire safety regulations because of the impact on 'UK plc'. That had paid for a test on the precise cladding product used on Grenfell Tower in 2001, saw that it failed devastatingly, sending 20m-high flames ripping through a test rig in just five minutes, but did not make a relatively simple tweak to official guidance to force it out of the market. And that wholly failed to implement the simple and clearly expressed recommendations the Lakanal House coroner made to prevent further deaths, with the official responsible for doing so writing that the government 'did not need to kiss her backside'.

The government was bound to an ideology that said it should not regulate the private sector, but should instead reduce any restrictions to allow it to generate economic growth. In the years before Grenfell, this became an all-out assault on regulation, codified in a 'red tape challenge' and a 'one in, three out' rule which effectively banned the intro-duction of any stringent new rules on business. In the crucial but niche area of fire safety in high-rise buildings, this utterly undermined its ability to ensure the safety of its citizens.

Throughout, the government relied on the fact that deaths in fires were falling to justify its failure to tighten fire safety rules. This is said to have been expressed by officials with the following phrase: 'Show me the bodies.'* There were simply not enough deaths to justify new restrictions on businesses. On 14 June 2017, our government got what it had asked for.

*

The Grenfell Tower fire killed seventy-two people, including eighteen children. It ripped families apart, traumatised an entire community, destroyed 129 homes and caused damage that, for many involved, can never be repaired. It is the most serious crime committed on British soil this century.

As I write, a police investigation remains underway, and a four-year public inquiry is reaching its final stages. That limits, to some extent, the conclusions that can be drawn from the evidence. Nonetheless, the evidence in itself paints a clear enough picture about the failures of the British state, the wrongdoings of various corporations and the incompetence of a string of public institutions.

It is a story about how these failures combined to inflict an appalling disaster on a west London community on 14 June 2017. Some families lost three generations in one night. Even those who survived the fire will carry the trauma of what they experienced forever.

---

* The witness statements of Sam Webb and Arnold Tarling said they had heard civil servant Brian Martin using this phrase. He denied having said it when asked at the inquiry.

Writing this book has always felt personal. On the morning of 14 June 2017, when I woke up to the images of Grenfell Tower on fire, my first thought was 'it's happened'.

I was working as news editor of *Inside Housing* magazine, a specialist publication for those who work in social housing, and in the months before we had written an ominous run of stories about fire safety. In March, my colleague Sophie had written about fears for other tower blocks, since some key changes recommended following the Lakanal House fire had not happened. Architect and fire safety expert Sam Webb told her 'really serious questions' should be asked in parliament about fire safety. He added that there was a 'conflict' between fire safety and the materials that are used to make buildings more energy efficient. He said: 'The materials used are not fire-resistant and in some cases they're flammable.'[3]

In April I had received a response to a Freedom of Information request I'd submitted regarding a fire in a block in Shepherd's Bush in west London the previous summer. It revealed the building had panels made of polystyrene and plywood fixed to its wall, which had given the fire a route up its external walls after a tumble dryer caught fire and flames burst out of the kitchen window. The experts we spoke to were clear: this could happen again. 'I'm worried about cladding systems in general. For a long time, people have been attaching things to the outside of buildings for insulation. It can be a catastrophic problem, particularly when flames can get in through windows and a "stay put" policy is in place,' Arnold Tarling, a chartered surveyor, was quoted as saying in the piece. We headlined the article 'A stark warning'.[4] I planned a broader investigation into cladding safety for the summer.

We'd also been working on a piece which suggested fire risk assessments were out of date in dozens of blocks around London. We knew residents would be told to stay put in a fire, and if the protections in the building failed, they would likely die. Risk assessments were all that was in place to ensure these protections remained adequate, but the data revealed they were not being done with any regularity. Our piece was underway, but unpublished at the time of the Grenfell fire.

A couple of days before the fire, I walked past firefighters attending a fire in a tower block near my home in east London. The road was cordoned off and residents in their pyjamas stood around on the pavement looking up at the building. I thought about a comment I'd heard several times from fire safety experts: if Lakanal had happened at night the death count would have quadrupled.

And then it happened. In the days that followed I asked myself if I'd done enough to sound the alarm. I've heard the same reflection since from dozens of people who had some inkling that a catastrophe was coming. But while none of us can go back, we can at least keep pressing for meaningful change. This is what I have tried to do in the five years since and this book is the culmination of that reporting. It is an enraging story, one which points to the deepest fault lines in our society. And if we want to repair them, it is one we must hear.

1

12.54 A.M.

In a kitchen on the fourth floor of Grenfell Tower, smoke began to seep out from the bottom of a fridge-freezer. All three people in the flat were asleep. A smoke alarm in the kitchen began beeping, waking the man on a mattress in the front room.

This man was Behailu Kebede, an Uber driver. He was the occupant of the flat, which he shared with two lodgers. He had returned from work at 11.30 p.m., showered, changed and fallen asleep on a mattress in the living room.

At the sound of the beeping, he went and looked in the kitchen. He was startled at the scene.

Behailu went back to the living room to call 999. While he was on the phone, he banged on the doors of the two lodgers to warn them of the fire. He got through to the emergency services.

'Fire Brigade,' the operator said.

'Yeah, hello, hi. In the fire is Flat 16, Grenfell Tower.'

'Sorry, a fire where?'

'Flat 16, Grenfell Tower. In the fridge.'

'The fire brigade are on their way. Are you outside?'

'Quick, quick, quick… It's burning.'

'They're on their way already.'[1]

Behailu was making the call from the lobby of the fourth floor and now began banging loudly on doors on the landing, rousing his neighbours in the six flats on the floor. His neighbours recall the bangs on their front doors were 'loud, mad knocking'. 'He [Behailu] was frantic, distraught and panicky,' recalls one.

Behailu dashed back to his flat and put on a pair of trousers. Reasoning that the fire may have been electrical, he switched off the red switch in his fuse box, cutting off the power to his flat. He then left his home and all his belongings: his black leather sofa, TV, fish tank, the picture of the Virgin Mary he had brought from Ethiopia. He would never return.

In the days to come, vicious media reports would carry the claim that he had packed a suitcase before leaving, with family members and friends pursued for comment and Behailu driven into hiding. The experience would have a severe and lasting impact on his mental health. These reports were entirely false. The inquiry praised Behailu's actions, saying he 'did exactly what a responsible person might be expected to do in the circumstances'.[2] The problem was that many, many others had not.

*

During the Grenfell Tower Inquiry, the company that made the fridge, Whirlpool, had challenged the conclusion that this is where the fire began by suggesting that (among other things) someone might have flicked a lit cigarette through the open kitchen window of Behailu's flat from the ground. The kitchen was on the fourth floor and this argument was dismissed as 'fanciful' by the inquiry chair, Sir Martin Moore-Bick, who followed two of his expert witnesses to the conclusion that the fridge was where the fire began.[3]

Behailu had bought this fridge five years previously in Brent Cross shopping centre for around £250. It had never been faulty before, apart from once when it iced over and had to be defrosted for a couple of days.[4] There was no reason to expect it to burst into flames, other than the fact perhaps that white goods occasionally do. The London Fire Brigade attends, on average, one such fire a day. According to a series of Freedom of Information requests made by the consumer magazine *Which?* kitchen appliances caused some sixteen thousand fires across the country between 2012 and 2018.[5]

An expert report would later suggest that a fault in the bottom of Behailu's fridge was to blame. The wires were improperly crimped, which made them vulnerable to overheating.[6] The expert claimed the wire overheated, causing a small fire in the box holding them which subsequently lit the plastic insulation backing the fridge. This conclusion was strongly challenged by Whirlpool in its evidence, and the final report did not make a definitive judgement about exactly how the fire began.

Nonetheless, the wiring wasn't the only issue with the fridge. Its plastic backing would not have been permitted in the US, where fridges are required to have a metallic casing. 'That steel backing would help to contain an internal fire, keeping it inside the unit, for a long time,' expert Dr John Glover told the inquiry.[7] A prior inquest in 2014, into a death in a house fire in Newcastle-upon-Tyne, had raised concern about the insulation used in a Hotpoint fridge-freezer, after 'very rapid' fire spread.[8] The fridge was a strange microcosm of the tower itself: clad in combustible plastic that let the flames spread despite prior warnings and tougher regulations elsewhere.

But at 12.54 a.m., this fire was still nothing unusual. It was the sort of incident that happens from time to time in a busy residential building. Blocks of flats should be built to withstand them. Grenfell Tower had almost certainly withstood such blazes and worse during its forty-year life. But now, things were different.

*

Behailu's 999 call had been put through to the London Fire Brigade's control room. Normally, this was based in Merton in south London, but on 14 June 2017 it had been moved to temporary premises in Stratford, east London.

The Stratford office is smaller and more cramped than the Merton base, with only sixteen desks for the operators, compared to twenty-nine in Merton. Merton also has two giant plasma screen TVs, which normally show a 24-hour news channel, and can be plugged into a feed from the police helicopter during an emergency. Stratford's single plasma screen was not working. It had one small television in the corner that could not be linked up to the helicopter and was not turned on.

At 12.54 a.m. when Behailu's call came in, it was a perfectly normal night. The night shift had been in since 8 p.m. and would clock off at 8 a.m. There were eleven staff on duty. The call was logged on the system and the computer automatically identified the three closest fire engines to send – two from North Kensington fire station and one from Kensington. A fourth was ordered minutes later from Hammersmith – a precaution given that the tower was a high-rise.

Over on the other side of the city from the control room, a siren went off in North Kensington fire station – raising the night shift from their sleep. The call sheet simply said Fire – Flat 16 Grenfell Tower, Lancaster West estate. Both of the station's fire engines were readied for action and ten firefighters climbed aboard – led by their watch manager Michael Dowden. Michael had been a firefighter for fourteen years. He had been along to Grenfell Tower before to familiarise himself with the layout of the tower and the estate but had never fought a fire in the building. He had no particular reason, at this point, to worry, and neither did the rest of his crew.

'From the information provided, there was nothing out of the ordinary and no cause for concern,' recalls one.[9] They expected to fight a simple flat fire and return to their beds in the station.

At this point they had limited information about the tower. The information in the brigade's database about Grenfell was many years out of date: it was incorrectly listed as a twenty-storey building, not twenty-four. There were no plans, no helpful photos. A box marked 'tactical plan' was simply blank and dated 30 October 2009. The inquiry report would later call the missing information for Grenfell Tower 'woefully inadequate' and 'inexcus-able'.[10] On the short drive, misgivings started creeping in for the crew. The screen in front of the cab also flashed up the information that the control room had now taken 'multiple calls' about the fire. 'I remember thinking "we have got something here",' recalls one of the crew.

I A.M.

In the tower, on the thirteenth floor, a student, Tiago Alves, was watching Netflix. He had lived in the flat since he was a baby, and been to primary and secondary school nearby. He loved his home and his community. Children from the estate attended his former schools. 'We'd always look out for one another,' he says. 'Everyone was extremely helpful, especially on the same floor. There was this sense of togetherness.' As a young boy he would play football with his friends in the lobby outside his flat – kicking the ball around the area outside the lifts or heading down to the football pitches next to the tower. 'To me it was my favourite place in the world,' he says.

That evening, his mum's cousin was in London visiting from South Africa. The family had eaten together at a res-taurant in Kensington Village, having talked about politics and the soaring house prices in the area before returning to the flat for coffee. At half past midnight, his parents had left to drive their relatives back to the hotel and Tiago had gone to watch Netflix in his room. It was his summer holiday from his second year of university and he was due to go on holiday to Switzerland with friends the next day.

At close to 1 a.m., his father burst into their flat. Tiago heard the door bang open and immediately knew there was something wrong – his father was always very careful to open the door gently to avoid disturbing the neighbours.

'Get dressed, there's a fire in the building,' he said. Tiago began pulling on his clothes, as his father woke up his sister Ines. She had an exam the next day – GCSE chemistry – and was annoyed to have been disturbed from her sleep. But Tiago's father had seen the smoke from the fourth floor on his way up the tower and he was adamant his family were getting out of the building. He had grown up in an area of Portugal where forest fires are common. From his perspective, if you were above the fire, you had to escape.

Tiago grabbed his phone, keys and wallet and left the building with Ines. His father, Miguel, stayed in the building knocking on all the doors on their floor to warn his neighbours to flee. Due in part to this early warning, everyone on the thirteenth floor would survive.

Tiago and Ines met their mother on the way down the stairs. She had seen the firefighters arriving and had let them into the building. The family left and gathered together with a small band of residents on the ground floor. After rousing the residents, Miguel also left the building. He gave his key fob to Michael Dowden to ensure fire-fighters could easily get back into the tower.

The fire engine from North Kensington was the first to arrive at Grenfell Tower at almost exactly 1 a.m. – six minutes after the 999 call. As Michael Dowden got out of the cab and put on his high-vis commander tabard, Behailu Kebede ran over. He told him the fridge was on fire and that everyone was out of the flat. From the ground, firefighters could see the orange glow of flames behind his window.

One team of firefighters were tasked with setting up a 'bridgehead' on the second floor of the tower. This is the base from which the firefighting operation inside the building is run. Michael Dowden stayed outside.

After being let into the building, the firefighters tried to use a special key to take control of one of the lifts. This is usually part of the strategy for high-rise fires: firefighters override the controls of the lifts to use for their operation. But it did not work. They inserted a key into the lock and turned it, but could not get control.

This would be a problem. Without an override, they could not take it out of service for residents, who may get trapped in it if the fire spread. They also could not guarantee its use for moving their heavy equipment up and down the tower quickly.

Why did the lift key not work? This question took up substantial time at the inquiry, and the evidence is not yet settled. The view of an expert witness is that the firefighters who arrived on the night used a key which did not fit.[11] But the Fire Brigades Union has strongly challenged the conclusion that this is what caused the lift to fail – pointing out that the switch itself was clogged with building debris left behind during the refurbishment of the tower.

One point worth making is that Grenfell had a complex 'drop key' mechanism for the lift override. A simpler option is available, more reliable and in use across much of Europe. The organisation that managed Grenfell Tower had chosen the more complex system because of a perceived risk of anti-social behaviour: they worried estate residents would buy skeleton lift keys online and deliberately take the lift out of service. They had also not fitted the lift with full firefighting features, including an escape hatch for firefighters, in part because of a perceived risk of anti-social behaviour, i.e. that residents might climb through the trap door and 'lift surf'. Prejudice against social housing residents appears to have actively undermined the safety features of the building.

Unable to gain control of the lift, one of the firefighters called the lift in the normal way and they went up to the second floor. A pair went up to the fourth floor, plugged their hose into the dry rising main and waited for it to fill with water. When the water came through, two more firefighters were sent up, wearing breathing apparatus and carrying a thermal imaging camera to fight the fire.

Standing outside the building, Behailu Kebede was filming the orange glow of the fire in the window of his kitchen on his mobile phone. It was getting brighter.

1.07 A.M.

The two firefighters wearing breathing apparatus sprayed a little water against the door to test its heat. If it sizzled and evaporated, they knew they would be immediately confronted by a hot fire. Seeing that it did not, they broke open the door to the flat with an enforcer.

They began to work their way through the flat – studying the thermal imaging camera to try and locate the fire. Hazy, grey smoke filled the flat. As they approached the kitchen, this smoke became thick and dark, the corridor was pitch-black and they were reliant on their thermal imaging equipment to see. As the smoke poured out onto the landing, the other firefighters retreated to the stairwell. One of them went up to the fifth floor to see if the fire had spread upwards.

Outside, the atmosphere among onlookers was calm. Parents were pointing out the flames behind the window to their young children and telling them that's why they should never play with fire. Behailu anxiously continued to film and Michael Dowden watched on. But the fire in the flat looked different. It appeared – somehow – to have breached through the window to the wall outside.

*

During the refurbishment, the tower had been fitted with new uPVC windows. These new frames were effectively glued on top of the old timber frames with a rubber membrane placed in between. The new windows were smaller than the old frames and needed to move out a few inches to sit within the new cladding system, which had also been installed on the external walls. This had created a gap between the old wall and the window, big enough to put your fist through. This gap had been filled with highly combustible insulation. In the top corner of the new window was a ventilation fan – sitting in a panel also stuffed with combustible insulation.

Much attention of three expert witnesses at the inquiry was devoted to precisely how the fire broke through this window. The eventual conclusion was that as the uPVC frame heated up, it began to melt and peeled away from the old wooden frame it was glued to, carrying the insulation board with it. The flames then simply burned through the rubber membrane and had an open route to the cladding system, the report said.* And this cladding system was extraordinarily combustible.

*

Shortly after 1 a.m., Rania Ibrahim made a video call to her sister, Sayeda. She told her there was a fire in the block and she was afraid she might not be able to leave. Her sister's daughter phoned 999 on her behalf and reported her location – but were told the fire brigade was already aware of the fire and to stay calm.

Rania lived on the twenty-third floor with her two children – Fethia and Hania, aged four and three. Her

* *Phase 1 report*, volume four, page 537. The inquiry report notes a separate simulation of the fire in Flat 16, carried out by the Building Research Establishment in May 2019, which suggested the more likely route for the fire's escape was the kitchen's extractor fan. A firm view is set to be reached in the *Phase 2 report*.

husband, the girls' father Hassan, was abroad attending to a seriously ill relative, and the three of them were alone in their flat. The family had moved into Grenfell Tower in February 2016. It had stunning views out across London. But it also made Rania afraid. She was claustrophobic and enclosed areas could provoke panic attacks. If there is ever a fire here, we will die, she had told another of her sisters, Rasha.

In fact, Rania had recently been plagued by fears of death. In March 2016, she visited Rasha in Egypt, where they had grown up, and told her she was having dreams about dying. But it was mostly a happy trip: Rasha was pregnant and Rania picked the name for her child, Sidra al-Muntaha, the name of a tree in heaven.

The two women had always been close. They grew up in the city of Aswan in southern Egypt and when Rania was born, Rasha was delighted to have a new sister in the family. Despite being a toddler, Rasha did not want anyone else in the family to hold the new baby. As teenagers, they had been inseparable – making regular trips to Cairo for Rasha to apply for a visa to join their older sister in the UK. 'We loved going to the cinema and buying clothes on our way. We shared too many funny things to mention in our travels,' Rasha recalled when I spoke to her in 2020.

After studying law at university, Rania had secured a visa to move to the UK and left Egypt in 2009. She married Hassan after meeting him at the local mosque and the family lived happily in west London, with many friends and relatives nearby. She continued to speak regularly to Rasha, who had stayed in Egypt, by video call.

But now she and her children were in danger. Her family waited, terrified, hoping the fire brigade would rescue her in time.

*

By 1.09 a.m. the fire had burst through the kitchen window and burning debris began tumbling down to the ground outside. The fire on the wall glowed white hot.

Michael Dowden told firefighters on the ground to direct their hoses above and below the window. On the advice of another watch manager present, Michael called for another two fire engines to attend.

The flat directly above Behailu's was Flat 26. A family of four lived there. At some time around 1 a.m., the mother woke up from where she had fallen asleep on the sofa watching TV. There was a ringing coming from the kitchen and when she went in, she realised it was the free carbon monoxide alarm the council had given them about three years previously. The kitchen was full of smoke.

But she was confused: nothing appeared to be burning. She walked around the kitchen trying to find the source of the fire. It was a hot night and the window was open.

'Within seconds of looking around the kitchen to try and find the source of the smoke, fire all of a sudden came through the kitchen window which was open and the blinds immediately caught fire,' she recalls. 'The blinds fell to the floor almost instantaneously. The colour of the flames was dark yellow/orange. The flames were covering the whole of the kitchen window which went up to the ceiling.'

The family fled, closing the door behind them. On the landing, they met the firefighter who had come to check the fifth floor. They told him their home was on fire.

Outside, the fire was now starting to climb up the building – burning up the shiny new cladding. 'As we watched I could see the flames were on the outside of the building,' says one witness. 'I was stood there in shock. I could not stop watching while the fire went higher up the building.'

Each floor in Grenfell Tower from the fourth floor up had six flats arranged around a central lobby, which contained the entrance to the lift and the single staircase. The layout was identical up the building – so as the flames rose from Behailu's kitchen in Flat 16, they began to attack the kitchen window of Flat 26 on the fifth floor, and then Flat 36 on the sixth and so on. The conversion of the bottom three floors of the building from commercial to nine residential flats had given the building its confusing numbering system: the flat numbers now corresponded to the floor three levels below. Above Behailu's flat were 114 homes. Many of their windows were dark. The tower's residents were sleeping, unaware of the danger growing below.

*

The cladding system on the walls of Grenfell Tower is known in the building industry as 'rainscreen'. This involves attaching insulation boards to the external walls of the building, leaving a small gap to allow moisture to evaporate and then fixing external cladding panels to a metal frame in front of it. It is designed to improve a building's appearance and insulation.

At Grenfell most of the insulation was made by a company called Celotex, out of the plastic polyisocyanurate. A smaller amount was made by their rivals Kingspan from a different plastic – phenolic foam. Both these plastics are flammable and emit toxic smoke when they burn.

The cladding panels were made from 'aluminium composite material', or ACM. ACM is effectively two thin sheets of aluminium held together by a plastic core. This makes the product more rigid, easier to cut and cheaper to produce than solid aluminium. But the downside is its fire performance.

The plastic bonding the metals together is polyethylene. It is made from petroleum, the same oil which powers our cars and warms our atmosphere, which is then converted into a resin, layered into sheets and set into plastic. It is literally solid petrol and will burn like it. In fact, in solid form it is denser and even more combustible.[12] By the time of the Grenfell fire, the world had seen ACM go up in flames – and send fire tearing up the side of high-rise buildings – several times before.

This is what was starting to happen at Grenfell Tower.

*

Inside Flat 16, the firefighters reached the kitchen. One of the firefighters put a pulse of water in from the hose. It immediately turned to steam: 'I thought "Wow". It was

hot and the steam cut through my PPE [personal protective equipment], I felt a burning sensation on my arms
from my elbow to my wrists, around the back of my neck
and head,' one recalls. It was completely black and the
thermal imaging camera was just showing white light.
They tried to find a new way in and approached the door
from the other side.

They did not know at this point that the fire now
burning on the outside of the building extended two
floors above.

Firefighters at the bridgehead on the second floor
encountered residents from these floors as they fled their
homes. 'Some people started to come down the stairs.
They said they were from floors five and six and that their
flats were on fire. They appeared to be overcome with
smoke, their eyes were streaming, they were coughing and
spluttering and looked panicked,' recalls one. 'I thought
they meant their flats were just filled with smoke rather
than being on fire… It would be very unlikely that the fire
would have licked up two floors. In my nineteen years in
the LFB I have only been to one fire where the fire has
licked up and caught the curtains, but never in a high-rise.'

The firefighters inside Behailu's flat were eventually able
to get into the kitchen, where one of the team was able
to 'knock it right out'. But the firefighters could still see
the blaze stretching up the walls of the building outside.
They desperately sprayed water onto the fire above them,
but to no avail.

'I remember the intensity of the flame,' recalls one. 'I
can only describe it as huge balls of flame falling down

along with debris; it didn't stop; it was violent… We kept hitting it with water but it was having no bearing on the fire.'

By now it was 1.20 a.m. It was twenty minutes since the firefighters had first arrived, and twenty-six minutes since the first 999 call. The firefighters had arrived promptly. Despite being delayed by the lift, they had got into the flat as quickly as they could. The inquiry report concluded that they 'acted as swiftly as they reasonably could'.[13] But the fire was already racing up the building. It was too late.

*

On the ground outside, Michael Dowden was starting to feel the pressure. He could see the fire getting worse. He recalls that as he watched, he saw the flames begin to take hold on the walls outside and begin 'spitting and sparking in a similar way to when magnesium burns'.

All glass windows will shatter and break eventually in a fire, and so once the blaze began to spread up through the cladding system it was inevitable that it would get into other flats. But the flaws which allowed it to break out through Behailu's window likely also aided its journey back into the homes of residents on higher floors. The uPVC frames would melt and deform, and the combustible insulation, rubber membrane, extractor fan and gaps in the system provided ample opportunities for the fire outside to get in.

In a flat on the seventh floor, a resident walked into his kitchen to find the extractor fan in the window burning.

It dropped into the kitchen, hanging by the wire that supported it, and orange flame shot through and set his curtains on fire. He pulled them off and stamped the fire out, but more flames breached through the left-hand side of the window and suddenly the entire kitchen window fell inwards. Black smoke rushed into the kitchen, stinking of plastic. He shut the door and fled.

Another resident recalls: 'As I got to the kitchen and looked down out of the window I saw a big fireball coming from the outside of the building. It was the colour of a burning sunset... The kitchen window then exploded inwards.' One more described the extractor fan falling in and black smoke shooting into the kitchen 'like water from a hose'.

Shortly before 1.20 a.m., Michael Dowden called for another two fire engines to attend. He was feeling increasingly overwhelmed: unaware of why the fire was progressing in the way it was, and without any real training or procedure as to how to fight it. The flames stretched up to seven flats above Behailu's, rising at a rate of just over one floor every minute.

'This is the point where I'm starting to become very consumed in terms of what was happening in front of me. I think the way it was increasing and developing, I've never seen anything like that before and it was almost that I was consumed by that in terms of the sensory overload,' he said. He was transfixed by the spread of the fire up the building but thought it was an external fire only. He had not appreciated that it would, by now, be breaking into flats and setting them alight. He did not know that firefighters inside the tower were encountering smoke above the fourth floor. He did not consider ordering the evacuation of the tower.[14]

The Grenfell Tower disaster was now underway. It had been more than forty years in the making.

2

'A DUMPING GROUND'

There are many points in history where the road to the Grenfell Tower fire could begin. But the one I will choose is 5.45 a.m. on 16 May 1968. On the eighteenth floor of a newly built tower block called Ronan Point in Canning Town, east London, a cake decorator, Ivy Hodge, lit a match to light her gas stove.

The resulting explosion blew her kitchen apart and caused one corner of the building to completely collapse. Four people were crushed and killed. Had it been later in the morning, or if the newly built block had been fully occupied, many more would have died.[1]

The flaws in Ronan Point were a product of politics. Amid a push to demolish and rebuild pre-war slums, successive governments competed to build more homes. Local government was encouraged and funded to build a new generation of social housing. And the encouragement was to build as high as possible: subsidies increased for every floor over five storeys. Between 1959 and 1967 an estimated 4,800 tower blocks were built. The new concrete skylines of our cities.

A new construction method had been discovered to build these tower blocks faster than ever before: large panel system building. This involved huge, preformed concrete slabs being driven to the site, craned on top of one another and fastened with bolts. It was rapid, efficient, modern and drastically reduced the number of builders needed on-site. The government got its housing numbers. The local authorities got their subsidies. Contractors and builders could make a lot more money building faster and cheaper.

But an awkward question was missing. Was it safe? The new buildings were like Jenga towers. If one block slipped, the entire structure could collapse. A young architect named Sam Webb began to fear a catastrophe. He had witnessed builders banging bolts flat with a sledgehammer because they didn't fit the concrete slabs. 'Nobody was really taking the safety issue seriously,' he recalls. 'It was huge money for them [the builders]. And everybody assumed that as the government was backing it, it had to be alright.'[2]

After one side of Ronan Point collapsed, investigations would reveal newspaper in joints that should have been filled with concrete, and rainwater seeping in between bolts. The links were weakened and when the gas cooker blew, the whole tower was ready to crumble.

But this was a political problem. What if all the blocks paid for by the government were unsafe? Where would the money come from to fix them? 'The main object [of the official response] was to keep politicians, of both parties, out of trouble,' says Sam.[3] Even Ronan Point was not evacuated. Instead, its collapsed side was rebuilt, strengthening work was carried out and residents were told to return to their homes or lose their social tenancies. Sam, who was by now campaigning on their behalf, predicted that it was a mistake.

He was right. In the mid-1980s, following safety campaigns by residents alarmed by the appearance of large cracks in the building, the chair of Newham Council's housing committee invited Sam to carry out checks on the building. He ripped a piece of wallpaper off a tenant's wall and dropped it through one of the cracks. It disappeared down the long hole in the building. 'One of the tenants said: "What have you done?"' he recalls. 'And I said: "I have just killed Ronan Point."'[4]

In 1984, the block was evacuated and its demolition ordered. But Sam wanted to see the true state of it, so it was taken apart carefully by his team of young architects and examined. 'I knew we were going to find bad workmanship – what surprised me was the sheer scale of it. Not a single joint was correct,' he recalled later.[5]

Following this, the housing minister in charge at the time said all blocks built in this way should be assessed.[6] But the government never took charge of ensuring this work was completed and for many, it would never be done.

While this saga was playing out on the UK's mainland, another disaster occurred on the Isle of Man. In August 1973, arsonists started a fire at an entertainment complex, Summerland, which had been built just two years previously. The blaze tore through the roof, made of combustible plastic, and the building was engulfed in flames with three thousand people inside. Fifty people were killed.

In these two disasters, we have the seeds of the Grenfell Tower fire. An out of control construction sector, a government unwilling to hear the truth about the risks, regulations failing to keep up with technology and the increasing use of combustible plastic as a building material.

These problems would never be addressed. In fact, new political forces were sweeping through the country which would only amplify them.

'WE NOW HAVE THE OPPORTUNITY TO MAKE RADICAL CHANGES'

Construction work on Grenfell Tower finished in 1974. At this time, builders in London worked under a set of codes called the 'London Model Byelaws', which effectively ruled out anything combustible on a building's external walls. This rule traced its history right back to the Great Fire of London in 1666 and a code that prohibited wood in favour of brick. 'Bricke is not onely more comely and durable but alsoe more safe against future perills of Fire,' the code said.[7]

This aversion to the use of combustible materials shaped London as a city of concrete and brick for three hundred years. I have heard one expert say it is part of the reason the city survived the incendiary bombing campaign by the Luftwaffe in World War II. But the old ways were being left behind.

In 1979, Margaret Thatcher led the Conservative Party to victory in the general election and embarked on a campaign to shrink the state. She is most famous for privatisation – British Gas, British Telecom, British Airways and a host of others were sold to private investors – but this was not all. Thatcher and her allies subscribed to an economic vision which said governments should avoid imposing rules on businesses and instead unleash markets to determine their own rules. She began a major campaign of deregulation.

This caught the construction sector in her second term. Michael Heseltine, then secretary of state with responsibility for housing, promised to deliver 'maximum self-regulation, minimum government interference' for the building industry.[8]

One arm of this was to part-privatise the enforcement of the rules. From now on, builders would be able to pay a private consultant, known as an Approved Inspector, to confirm their projects had complied with building regulations, instead of being required to seek sign off from the local council.

But the rules themselves were also changing. Designers and builders were hungry to be let loose to 'innovate': modern buildings, like new shopping malls, did not fit in well with old, prescriptive rules around – for example – the minimum time required to reach a fire exit. In a debate in parliament, one supportive MP said the aim was to reduce the 'petty frustrations' of big builders and 'help keep costs down'. Mr Heseltine put it as follows: 'The present system is far from ideal. Builders complain of the delays and costs which it imposes. Designers object to the limits that are imposed... We now have the opportunity to make radical changes.'[9]

In 1984, a new bill was introduced which aimed to sweep an estimated 350 pages of previously existing building regulations into the bin, and replace them with just 24 headline standards. All local and regional codes would be replaced – including the tough restrictions relating to external fire spread enforced by the London model byelaws.

The new regime shifted from being 'prescriptive' to 'performance-based'. This meant rather than being given a book of rules to follow, builders would simply be told to achieve certain outcomes and left to decide how to do it themselves. For external fire spread, the new rule became: 'The external walls of the building shall resist the spread of fire over the walls and from one building to another, having regard to the height, use and position of the building.'

Alongside these 'performance-based' regulations, the government introduced new, non-mandatory 'approved documents' containing official guidance about how to meet the standards. The rules about fire were contained in Approved Document B.

There were those who issued warnings. As the act neared Royal Assent, Labour peer Lord Sydney Irving tried to sound an alarm: 'I hope that when the government come to consider the building regulations and the guidance in the approved documents they will not permit any relaxation in such matters as non-combustibility and remember that on such occasions when building standards have been relaxed it has often led to disaster.'[10]

His warnings fell on deaf ears.

'WE HAVE RECEIVED...A REQUEST FROM MARSHAM STREET PRESS OFFICE TO PLAY DOWN THE ISSUE OF THE FIRE'

The concrete towers built with such enthusiasm after World War II were becoming increasingly uncomfortable by the 1980s. Many were draughty, cold and expensive to heat.

A new means of improving insulation was being developed: overcladding. This meant adding a system to the external walls of the block which would improve its thermal performance. But this could go wrong. As early as 1986, the government was aware that doing it carried a potential fire risk. 'A risk of increased vertical fire spread has been identified during the laboratory testing of over-cladding systems incorporating combustible insulants,' said a circular issued by the Department of the Environment (DoE) on 9 December 1986. 'These emergent flames could re-enter the block via windows.'[11]

Nonetheless, it was keen to investigate the use of such systems. In 1989, the government committed almost £1m to install a pilot cladding system on Knowsley Heights, an eleven-storey block in Merseyside. The work was to be monitored by the Building Research Establishment (BRE) – then a state-owned testing and research facility for the built environment. The tower was fitted with a 'rainscreen' cladding system. Rolls of non-combustible insulation were fitted flush to the concrete walls. Then panels made of glass-reinforced plastic were fitted to an aluminium frame in front of this insulation. A similar system would be used on Grenfell Tower twenty-five years later.

The project was completed in 1991 and appeared to be a success. 'The techniques and philosophy employed at Knowsley could be readily adapted to other multi-storey dwellings,' said the BRE report. But something was about to go wrong.

At 2 a.m. on Friday 5 April 1991, arsonists started a fire in bins outside the block. The fire ripped up the building – breaking into flats on all eleven floors and racing right up to the roof by the time firefighters arrived. 'It was the most frightening thing any of us had ever seen as firefighters,' said one of those who attended. 'Flames were coming from every landing window between the ground floor and the roof.'[12] The BRE was instructed to produce a report. But it appears someone within government didn't want the full story being made public.

A handwritten note, undated and signed only by 'Lyn', which I unearthed from the National Archives documents on the fire, reads: 'We have received via HMEA a request from M St Press Office to play down the issue of the fire. Our briefing to the secretary of state will be purely factual and as far as we are aware Knowsley [Council, the landlord of the block] will not be making an issue of the fire.'

'M Street Press Office' appears to be a reference to Marsham Street – where the government department sponsoring the project was based. HMEA is the acronym for Housing Management Estates Action, the government team which administered the funding.

An internal report written by an HMEA official days later said the fire was 'being treated as insignificant'. 'The block performed well, the fire doors were un-vandalised and in place (which is unusual for Merseyside), the building was evacuated easily, only three people were affected by smoke and all tenants returned willingly to their flats,' the official wrote.

Why did the government want to play the fire down? A handwritten letter from the archives may hold the answer. It says the project 'is of particular interest to the department in that Knowsley Heights... was overclad using techniques relatively new to public sector housing in this country but which are being replicated on other blocks'. Reference was made to a project in Lambeth which was 'continuing to proceed' despite the fire.

But if this fire had been properly scrutinised, it would have demonstrated the dire need for change. The combustible cladding panel on the external walls had a fire rating of Class 0. And this meant it complied with the standards in the new Approved Document B.

The Class 0 rating is crucial to the Grenfell Tower story. It is derived from two tests contained in an old British Standard document, and primarily assesses the spread of flame over the surface of a material. But this makes it flawed. If a material is packed with fire-retardant chemicals, or if it is a composite – made of one material on its surface but different throughout – a material could pass the test in a laboratory, but still be extraordinarily dangerous in a fire when bolted to the external wall of a building.

A fire at Knowsley Heights tearing through Class 0-rated cladding in 1991 was the first major opportunity to realise that the guidance was seriously, even fatally, flawed. Class 0 could and should have been binned in favour of a tougher standard for cladding.

But the BRE's report drew no particular attention to this flawed standard. In fact, it suggested the cladding system used had been non-combustible and pointed to missing fire barriers as the primary problem, not the combustible cladding panels. The report concluded that there was 'no reason to suggest a life risk associated with cladding' unless there were large gaps between the cladding and insulation.

In the years which followed, Knowsley was misinterpreted as a cautionary tale about missing fire barriers – not the need to avoid using combustible cladding. Had it not been 'played down', the entire crisis may never have happened. 'I think it's impossible to overstate the importance of what was missed here,' expert witness Professor Luke Bisby told the inquiry in summer 2022.[13]

It would not be the last missed opportunity.

'THIS SERVES TO UNDERMINE THE INTEGRITY OF THE REGULATIONS AND THEREFORE REDUCES FIRE SAFETY'

Following the Knowsley Heights fire, the government asked the BRE to work on a new 'large-scale' cladding test to provide a more sophisticated means of assessing the fire risk from cladding systems. This work resulted in the construction of a four-storey test rig at the BRE's headquarters in Cardington. Mock tests were carried out and a report was written in 1994. This research plainly identified the problem with Class 0 cladding panels. 'It is clear… that a cladding material achieving a Class 0 rating may suffer extensive surface burning. This burning often spread to the top of the test building (some 9m) and would have spread further if possible,' it said.[14] But still no changes were made to guidance.

In June 1999, another fire struck – this time at Garnock Court, in Irvine, Scotland. Flames ripped up a strip of cladding which had been installed beneath windows in 1991. The fire spread from the fifth floor, and tore all the way up to the roof. Flames and dense black smoke spread into the sky. Firefighters struggled to deal with fierce, simultaneous fires on nine floors. Pensioner William Linton was trapped on an upper floor and killed.

The BRE was asked to prepare a report into this fire – one for central government and one for North Ayrshire Council. It found that once more, glass reinforced panels were involved. 'The plume will have ignited the GRP [Glass Reinforced Plastic] and remained in contact with it and generated a self-propagating fire,' it said.[15] In the report the BRE prepared for North Ayrshire Council, it noted that these panels were required to be Class 0 and recommended that when the building was refurbished, the council should use 'non-combustible' materials. But curiously, these passages were omitted from the report returned to government. No BRE or government witnesses were able to explain why when asked about this at the inquiry.

These two fires did not go unnoticed in Westminster. A committee of MPs set up an investigation into the risk from cladding fires. They called evidence from the BRE, the department and a variety of industry experts.

Here it is important to note another fire classification: limited combustibility. This is a much tougher test than Class 0 and effectively rules out any combustible products like plastic or timber. In 1999, a rainscreen cladding system was required to have limited combustibility insulation, but only a Class 0 cladding panel. Dr Bob Moore, an expert from the Fire Safety Development Group, which represented the makers of fire safety products, wrote a report explaining this problem to the committee. 'Combustible materials, like plastic, wood, etc are *not* materials of limited combustibility but can achieve Class 0 performance by adding fire-retardant chemicals or facing the combustible material with a metal foil or sheet... This serves to undermine the integrity of the regulations and therefore reduces fire safety,' he wrote.[16]

But the voice of the cladding industry urged caution. 'Any changes to the facade to satisfy a single requirement such as fire performance will impinge on the wall's performance as well as its cost,' said its industry representative, adding that fire-resistant external walls were not 'economically viable'.[17]

In the event, the MPs, in a report issued in December 1999, told ministers to scrap the Class 0 standard and require all cladding systems to either be entirely non-combustible or able to pass one of the BRE's large-scale tests.

They also called on the government to instruct housing providers to check the safety of any cladding systems currently installed on their blocks, and to include them in risk assessments going forward. 'We do not believe that it should take a serious fire in which many people are killed before all reasonable steps are taken towards minimising the risks,' their report said.[18] But the government did not follow this advice. In April 2000, it elected to stick with the Class 0 standard and simply introduce the BRE's large-scale test as an alternative. Why? A former official told the Grenfell inquiry that the requirement for fully non-combustible cladding systems was considered 'impracticable and unduly onerous', while asking for large-scale tests of all systems would have been 'unpopular with industry'.[19]

Dr Bob Moore spoke to me in 2021. 'They just didn't think it was worth doing,' he said. He retired shortly afterwards and did not think about the issue again, until he saw the Grenfell Tower fire on television in 2017. 'I always wondered if I was to blame,' he said. 'Perhaps I should have said more.'[20]

'IF THIS OCCURS THERE COULD BE ECONOMIC CONSEQUENCES FOR THE BUILDING INDUSTRY AND THE UK AS A WHOLE'

In 1999, Brian Martin took a job at the BRE. Previously, he'd been a building control surveyor, working for local authorities to check and sign off plans submitted by developers. He'd taken an interest in the fire safety elements of the job, but had no formal qualifications in fire safety issues. Before becoming a building control surveyor, he'd worked in the construction industry as a joiner and site manager – where his experience of fire safety matters amounted to 'fitting a few fire doors'.[21]

Nonetheless, he was about to take on an important role. The BRE had been privatised in 1997, ending almost eighty years as an arm of the British state. It continued to advise government, but now through formal, private contracts with central government. One of these was to support its work maintaining and updating Approved Document B and within a few weeks of starting at the BRE, Brian was seconded to Whitehall for two to three days a week to advise on the guidance under this contract.

And the guidance was about to change. While the government had elected not to follow the select committee's advice on Class 0, it had agreed to introduce the new large-scale test designed by the BRE following the Knowsley Heights fire as an alternative way to assess the safety of cladding systems. The BRE was contracted by government to finalise this test: to determine what the appropriate 'pass/fail' criteria should be, and to examine how the widely-used cladding systems fared.

The BRE was supposed to carry out a 'comprehensive' survey of cladding systems in use in the UK to support this work, but when it surveyed local authorities it was only able to obtain responses from thirteen – a tiny fraction of over three hundred around the country. It also took data from industry journals to understand what cladding products were on the market.

This resulted in fourteen cladding systems being designed, with the tests run between May and November 2001. But one test, carried out on 18 July 2001, stood out.

The BRE tested a cladding system with non-combustible glass wool insulation and a cladding panel made of ACM. The results were astounding.

After just three minutes, the failure criteria had been exceeded. And in less than six minutes, the flames from the burning cladding extended 20m into the air – more than double the height of the 9m rig. The test was halted and extinguished to protect the safety of those present. The risk of using polyethylene-cored ACM cladding, which would become so devastatingly apparent sixteen years later, was revealed in British-state funded testing in the summer of 2001.

What was even more worrying is that the ACM cladding panel used in the test had obtained Class 0 and could be used on tall buildings in compliance with the guidance in Approved Document B. Change was urgently needed to prevent a catastrophe. But in the report it submitted to government in September 2002, the BRE did not spell this out. Its report noted that the material had satisfied Class 0, but had still 'proved to be one of the worst performing products' in the large-scale test. But it merely said that 'these issues require further consideration.'*

There was industry pressure not to impose tougher standards. The CWCT, which represents the cladding industry, warned in August 2002 that the new testing methodology, which had seen a number of other cladding products fail, could result in 'the abandonment' of the use of rainscreen cladding systems. 'If this occurs there could be economic consequences for the building industry and the UK as a whole,' it said.[22]

When asked by the inquiry, former official Anthony Burd said he felt it was unnecessary to change the guidance. The statutory requirement that buildings should 'adequately resist' the spread of flames meant highly combustible products like ACM should not have been used – even if they obtained Class 0. But this logic felt circular. The evidence that they were so dangerously flammable came from the testing. And the testing was never released. Asked bluntly at the inquiry if this was a 'cover up', Mr Burd insisted it was not.[23]

Even without this testing, the government should have dropped Class 0, simply to harmonise its standards with those in the European Union. The EU had adopted a new

---

* Colwell, S. 'Analysis of ISO 9705, European and British Standard fire test data for BR135 project.' When asked why they did not explain more clearly that Approved Document B needed to be amended as a result, BRE witnesses said they would not give direct policy advice in this way.

testing methodology which graded building products alphabetically from 'A1' (the best) to 'E' (the worst). The UK, as a member state, was supposed to adopt these standards and ditch its parochial national classifications to ensure consistent quality across the bloc. Minutes of a meeting discussing the change show officials were warned the UK faced becoming a 'dumping ground' for inferior products if it did not adopt the higher standards.[24]

Officials elected to use Euroclass B as the alternative to Class 0. This was lower than many European states and would still permit the use of some combustible materials. A research report prepared for the government in May 2000 warned that setting higher standards would 'severely restrict market choice'.[25]

But Euroclass B was at least tougher than the defunct Class 0. This new European standard should have come into force by 2005, replacing the outdated national standard. But it did not. Instead, the country entered a transition period where both Class 0 and Euroclass B could be used as the standard for cladding panels, a position which continued until the Grenfell Tower fire. Documents suggest this was because of industry lobbying. Manufacturers feared materials which could achieve Class 0 would only have obtained a 'C' or 'D' under European standards and their representatives lobbied against the change. In May 2003, a pamphlet produced by insulation manufacturer Kingspan – who we will meet in later chapters – said: 'Government has stated that it will not implement the new Euroclass system until industry is ready for it.'[26] It appears they never were.

'THE INTENT OF THE ORIGINAL GUIDANCE HAS BEEN CIRCUMVENTED'

Another huge tower block fire struck the UK in June 2005. This time, flames ripped up from the second floor to the top of The Edge, a nineteen-storey block in Salford, in less than ten minutes. Investigations revealed the building had an exterior made of 'sandwich panels' – metal sheets held together with highly combustible polystyrene insulation. Once more, the problem was obvious: the panels had a Class 0 rating meaning they had been used in compliance with Approved Document B.

This fire should have come at the ideal moment to result in a change to guidance. A new version of Approved Document B was due to be published in 2006 and Brian Martin and his colleagues were working on updating it. Surely, with the risks of combustible cladding made obvious by yet another real-world fire, the Class 0 standard would finally be dropped?

In January 2005, Brian Martin wrote to colleagues, citing The Edge fire, to say that the 'guidance is in need of clarification'. 'This appears to be as a result of changes in construction practice whereby the intent of the original guidance has been circumvented,' he added.[27]

He suggested changing the wording of the passage about 'insulation' to say that it also required 'any other material' used in an external wall to meet the tougher standard of 'limited combustibility'. This would have ruled out combustible cladding products like ACM at a stroke. But that is not what happened. Mr Martin was concerned this change would have 'prohibited timber-framed buildings' – where buildings are built with a frame made of wood instead of the more traditional masonry or steel.

Mr Martin's solution was to amend a paragraph headed 'insulation materials/products' to say that any 'filler material' should meet the standard of limited combustibility. He hoped this was a neat solution – capturing the core of a sandwich panel like the one used at The Edge, while allowing timber construction and certain other combustible products to continue being used.

A huge amount now hung on these two words. It was the only indication that materials like the ones used on The Edge and in the terrifying 2001 test were banned. The prevention of a future tragedy depended on the industry realising what was meant by the words 'filler material' and no longer installing these dangerous materials on people's homes.

The trouble was that it was not at all clear what Mr Martin meant by the words 'filler material'. The clause was contained within a paragraph headed 'insulation materials/products', but an external cladding panel is not insulation. And to many in the industry, the phrase 'filler' implied a Polyfilla-type product used to plug up gaps.

Martin and his team never consulted on the meaning of the word 'filler'. When they sent out a circular summarising the key changes, they did not mention it. Asked at the inquiry in 2022 if he was trying to 'slip the change in under the radar' to prevent industry challenging it, Martin accepted this was part of the thinking.[28]

But meanwhile, the Class 0 rating remained in the guidance. And with nothing to tell builders that this no longer covered the inside of a sandwich panel, they went on installing combustible products, seemingly paying little regard to the overall legal requirement that buildings should resist the spread of flame. As refurbishments of the high-rise council homes built in the aftermath of World War II were carried out, more and more combustible materials were being added to their walls. These blocks were now almost half a century old. They were beginning to age, and were becoming less fire safe as they did so. The fire risk was starting to feel like a perfect storm. And on a hot day in south London in early July 2009, that storm would break.

'OH MY GOD, NO, LISTEN, I CAN SEE FLAMES AT THE DOOR'

Lakanal House is a fourteen-storey block of council housing, built in the 1950s, comprising ninety-eight maisonette flats, each spread across two floors. The building was named after Joseph Lakanal, the founder of the French education system, with all blocks on the surrounding estate named after prominent French figures, for reasons best known to the original planners.

In 2006–7 panels were added to the outside of the flats beneath windows for aesthetic and insulation purposes in a major refurbishment. They were made of a material called 'high pressure laminate', essentially compressed layers of wood and glue, which sandwiched together combustible insulation. These, combined with serious flaws inside the building, meant it was now a fire trap.

It was just after 4 p.m. when Catherine Hickman began to smell smoke. She was at home alone – her boyfriend Mark, a successful hair stylist who worked with celebrities, was in New York.

Catherine's star was on the rise: a fashion label she'd established called Moi et Cat had clothes stocked in trendy outlets in the USA and Japan. The singer Björk had commissioned the label to produce outfits for her, and her designs appeared on catwalks from New York to Reykjavík. 'The path she had followed with her chosen career had finally paid off and she was embarking on a fabulous adventure in fashion,' her family would later say. In 2007, she'd moved into the maisonette in Lakanal House with Mark.

A couple of weeks before, she had spoken to him about her fears over fire safety. 'Shouldn't there be a plaque or a poster on the wall showing an escape route or what we do in case of a fire?' she'd asked.

At 4.21 p.m., she looked out of her window and saw flames coming out of the window of the maisonette below. She dialled 999.

'I'm at flat 79 and the flat below me, there's flames coming out of – coming out the window,' she said.

'Right, okay. You need to stay in your – in your flat,' the call handler replied.

Catherine did as she was told. But the conditions started to get worse. Flames were licking up the outside of the building. The external panels on the outside of the building were burning. Soon, the flames buckled her window frame, smashed the glass in her window and set light to her curtains.

Catherine stayed on the phone to the emergency services. She asked if she should leave but the call handler advised her to try and stop the smoke coming in through the door. Still on the phone, she fled to the upper floor of the maisonette.

'There's fire coming through my floorboards now, and smoke,' she said. 'What shall I do? Shall I get out?' The operator told her to go into a room where there was less smoke. 'I want you to stay there. I don't want you to try and move, okay?' the operator said.

The flames continued to close in. 'Oh my god, no, listen, I can see flames at the door,' she said. At 4.45 p.m., she suddenly screamed. 'Something hot fell on me. It's falling from the ceiling.' The operator advised her to crawl to a different part of the smoke-filled room. Her voice became fainter. Four minutes later, twenty-eight minutes after the call began, she stopped responding completely.

Outside the building there was pandemonium. Residents were fleeing, firefighters were arriving and police were trying to prevent worried onlookers and relatives from getting into the building. The flames from Flat 65 were ripping up the building in a zigzag. Black smoke was pouring into the clear blue London sky. And outside Rafael Cervi was trying to get into the building to find his wife Dayana Francisquini and their children Thais and Felipe.

Rafael and Dayana had moved into Lakanal House in 2006. They'd met in a Brazillian club in 2003. He was introduced to her daughter, Thais, who was three months old at the time. 'After that, I could not leave both of them,' he recalled at the inquest in 2013. The couple moved in together and a couple of years later, in 2005, their son Felipe was born. Thais, who Rafael regarded as his own daughter, was a model schoolgirl. 'Everything was always lovely with her. She got plenty energy, very clever kid,' Rafael said. He had just started teaching her to ride a bike. Felipe, aged three, was mischievous and 'always doing crazy things' like stealing hair gel and covering himself in it. Rafael's mum said he took after his father.

On the day of the fire, Rafael had been working near London Bridge – not far away from Lakanal House. At around 4.30 p.m., he noticed a missed call from his wife. When he called her back, she told him the building was on fire and there was smoke coming into their flat. She said she was sheltering in the bathroom, where the smoke was less thick. He phoned the fire brigade at 4.42 p.m. to tell them where she was. Panicking, he began rushing back home. When he arrived and saw the fire, he tried to get into the building but was stopped by the police. He gave the location of his family to the firefighters and waited desperately, phoning Dayana again and again to find out what was going on. She told him the family was trapped in the bathroom. The conditions were bad and Felipe was struggling to breathe.

Next door Helen Udoaka was also being affected by the smoke. She was at home with her baby, Michelle, who was just twenty days old. Her husband Mbet was at work. The couple had met in Nigeria in 2003 while studying at the University of Lagos. Mbet had moved to England first, and Helen had joined him in 2007, moving into his flat in Lakanal House. He was working part-time in security while doing a business course. She was studying an NVQ in care and social services. They were thrilled when they found out they would be having a baby girl.

Like Rafael, Mbet got a phone call at around 4.30 p.m. from his wife telling him there was a fire in the building. She said their flat was full of smoke and she couldn't breathe. He also got into a cab and rushed to Lakanal House. During the journey Helen told him she had moved – and was now in the bathroom of Flat 81 with Dayana, her children and another family – who eventually left the bathroom for the balcony and were rescued. But Helen and Dayana stayed – waiting for the fire service to save them. They never did. At 5.38 p.m., Dayana made her final call to the emergency services. Five minutes later, Helen made hers. The bodies of the two women and the three children were found later that evening.

'That afternoon I saw everything that I build, everything that I dream of was over,' Rafael told the inquest into their deaths four years later.

Helen's father, in his seventies, died of a heart attack the night Mbet told him what had happened to his daughter and granddaughter. 'My life will never be the same again, and I can never get over these deaths,' Mbet told the inquest. 'This is what I keep saying to Helen each time I visit the cemetery, because the cemetery wasn't what we planned.'

The disaster proved the danger which had been mounting in high-rise buildings since the 1980s. It was up to the British state to ensure it was never, ever repeated.

3

1.20 A.M.

In Stratford, the control room was starting to appreciate the scale of the fire they were dealing with. And at 1.21 a.m., they received their first call from a resident inside the building, other than Behailu, who first reported the fire, and had confirmed to the brigade that he was now safely outside the tower. A woman on the twenty-second floor phoned to tell them she could smell smoke. She was told to close her flat door and stay where she was.

A few minutes later, in the words of one operator, 'all hell broke loose'. Between 1.24 a.m. and 1.26 a.m., the number of calls from the incident rose from nine to seventeen. By 1.30 a.m., it was twenty-nine. Phones were ringing constantly, and it was now obvious that the operators were dealing with a major and rapidly worsening fire. One recalled thinking 'Oh my God, this is worse than Lakanal.'

At this point, the control room officers were not seeing what those on the ground were seeing. On their screens they had a message saying a fourth-floor flat was 75% alight. It was not updated with the fact that the flames were tearing their way up the outside of the building.

While they were getting individual reports from callers describing 'a whole tower block on fire' and 'a line of fire going right up the outside of the tower', no one was putting these together to form an overall picture and realise what was happening on the ground.

As such, no one queried the orthodox London Fire Brigade policy for high-rise buildings: tell callers to stay put. Residents who called from inside the tower were told to shut their doors and windows and stay where they were.

At Lakanal, callers had also been told to stay put and had lost their chance to escape as they awaited rescue. As a result, the fire brigade had told the coroner investigating the six deaths at Lakanal that it had improved training to call handlers as a result for circumstances known as 'fire survival guidance' calls.

But, as we will see in Chapters 16 and 18, far too little had in fact changed: stay put remained the default position. Eight years on from the tragedy of Lakanal, many of the same mistakes were about to be made again – on a much larger scale.

*

Outside the building, incident commander Michael Dowden was desperately trying to work out a strategy. The blaze was now tearing up the building. Giant sheets of metal up to a metre square were peeling off the 70m-tall building and spiralling to the ground. 'Some of the debris was molten and as it fell to the ground it was obviously still alight,' he recalled in his witness statement.

At 1.24 a.m. he called for ten fire engines and by 1.27 a.m., with the fire showing no signs of abating, he called for five more including two with long ladders. He was now seeing people leaving the building suffering from smoke inhalation and radioed through a message saying the fire was now 'persons reported'. This was a significant message – it meant people were affected by the fire and in danger. Two minutes later, on the advice of another watch manager who had just arrived, he called for five more fire engines. The plan was still to fight the fire and bring it under control. A crew was deployed to try and reach the roof to fight the fire from above.

'There were probably moments when I did feel helpless,' he would later tell the inquiry. 'It's a very, very difficult position to be as an incident commander when it's just, it's just relentless… This was like nothing else I have ever experienced before. The ferocity, the way that fire was developing was just relentless.'

At around this time the first police officers arrived on scene. 'Other flats at risk of fire, going to be a massive evacuation,' one said in a radio message to his base at 1.23 a.m. But this was not what Michael Dowden was planning. He was continuing to operate under the belief that the firefighters could bring the blaze under control and did not know it was penetrating the building internally. He gave no real consideration to evacuating the entire building.

*

Why did Michael Dowden not order an evacuation? At this point, residents were able to walk down the stairs and out of the building relatively untroubled. The stairs remained clear of smoke, and so were many of the lobbies. In the fifteen minutes between 1.15 a.m. and 1.31 a.m., seventy-seven people walked down the stairs and left the tower without injury. One expert estimated (not taking into account age and vulnerability) that it would have taken all 293 people present on the night of the fire seven minutes to walk down the stairs and exit the building, if there had been some means of alerting them all to escape.[1]

Yet the UK's reliance on stay put meant that Grenfell, like almost every other high-rise building in the country, was not fitted with a communal fire alarm. In reality, Dowden's only options would have been to instruct his crew to shout from outside with loud hailers, tell call handlers to warn any callers who phoned 999 to leave and to send firefighters door-to-door telling residents to get out and assisting them where necessary.

But this represents a complicated operation with serious potential risks, and Dowden had never had a scrap of training on how to implement such an evacuation. He had never been trained on when to drop 'stay put', or how to evacuate a building. Neither had the rest of the crew who arrived from North Kensington fire station, despite a combined fifty-two years' service.

'As a firefighter, you're told to go by the book. You aren't supposed to improvise,' one experienced firefighter told me, after watching Michael Dowden questioned at the inquiry. 'You need to know that your mate is going to do what you expect him to do, because in the middle of a fire your life depends on him following the same plan you are. If Michael Dowden had improvised an evacuation and thirty people had died, he'd have been the firefighter that killed thirty people.'[2]

The London Fire Brigade had only ever taught its incident commanders to rely on stay put. We will come back to the reasons why in Chapter 16, but this was a situation severely criticised by Sir Martin Moore-Bick in his final report. He said the concept had become 'an article of faith within the LFB so powerful that to depart from it was to all intents and purposes unthinkable'.[3]

But inside the building, many residents were fleeing the tower on their own initiative. Those in the 'flat 6s' were seeing fire break into their kitchens. As it did, they left – often in a state of panic. One woman recalls being awoken by the commotion outside the building. When she went into the kitchen, she saw some sparks in the air. Wires around the kitchen fan started to burn. She cleared things away from the window, hoping the fire would stay out of their flat. But then the window broke. 'The whole window in the kitchen broke in two,' she says. 'The vent completely collapsed. I saw the glass break and the fire come through as I was standing there. The plastic around the window was burning.'

The instinct of most of these residents was to get out of the building. With the lift still in service a group of five residents entered it. As they descended, the lift stopped unexpectedly on the tenth floor, the doors opened and thick black smoke poured in. 'It was terrifying, and the smoke was horrible. There was a strong and bitter chemical smell. I tried to speak, but it was too hard because of the smoke,' says one passenger. 'I tried to get out but it was difficult. I realised there was someone behind me who was holding onto me. I couldn't see who it was because it was so dark.'

Three of these residents exited the lift amid the panic. The smoke conditions on this floor were assessed as so severe that they could induce collapse within thirty seconds to two minutes. They would never find the exit to the stairs in the dark, choking conditions. Their bodies would be carried out by firefighters hours later. They are likely to have been the first victims of the fire.

*

Natasha Elcock, a supermarket manager who had lived on the eleventh floor of Grenfell Tower for twenty-one years, had fallen asleep on her sofa with her partner watching *Trainspotting*. She had been woken up just after 1 a.m. by voices outside her flat. She opened the door and found her neighbours on the landing, who told her there was smoke in their flat.

Going back into her home, she and her partner looked out of the window. They could see the firefighters below, spraying water onto the column of flames stretching up from the fourth floor. Natasha got dressed and phoned 999 at 1.28 a.m. She said she was on the eleventh floor and didn't know how to get out. She said she had seen smoke on the landing but not in her flat. She was told to keep the door closed and stay where she was. Firefighters would be given her location, the operator said. Natasha was not particularly worried at this stage. She knew there had been fires in other buildings locally which the fire brigade had dealt with. She had faith that they would put this one out too.

By 1.27 a.m., the flames reached the very top of the tower. The images were eerily similar to those of other ACM fires in other countries – a vertical column of white-hot flame. But this blaze had not finished. It was about to get far worse than anything seen before.

4

'SHOW ME THE BODIES'

Following the Lakanal House fire in July 2009, the government asked the BRE to investigate what had happened. The organisation had an ongoing contract to examine the consequences of other significant fires, and whether they revealed anything which merited altering building regulations.

The BRE's investigators attended the site shortly after the fire and made some initial investigations. Blocks of flats should resist the spread of flame, allowing the firefighters ample time to extinguish the fire before it uncontrollably spread. Either the building did not comply with the regulations, or the regulations were not sufficient, but something had gone wrong. It was the BRE's task to find out what.

But within government, it appears minds were already being made up. On 14 July – before any report from the BRE – Brian Martin, the former joiner who was now the civil servant effectively in charge of fire safety regulations, emailed a fire engineer who had inquired about the blaze to tell him: 'Based on the snippets of info I've had so far. I don't think there's any need for changes to [Approved Document B].'[1]

Two days later, the BRE delivered its initial report. This report was able to identify the presence of plastic panels on the walls of the building, and that fire spread externally. There was plainly more work to do to understand why this happened and join the dots to ensure it did not happen again.

But the BRE was not asked to do anything more. On 28 July, before it had submitted its final report and with its investigations still at an early stage, its investigation was shut down. 'For the purposes of the fire investigation contract you have with the department, I'm satisfied that there will be no need for you to revisit Lakanal House,' Mr Martin wrote. 'Any further visits will need to be funded by a third party.'*

Internally – according to its witnesses at the inquiry – the government was already coming to the view that Lakanal House was a case of non-compliance. There was no need to amend the guidance or impose tougher regulations on the use of combustible materials on high-rise buildings.

This view was tested in December 2009. The Metropolitan Police and London Fire Brigade had continued to probe the causes of the fire. In November, they had asked the BRE to run tests on the panels used on the walls and found they burned fiercely and could not even obtain the limited Class 0 classification.

---

* Inquiry transcript, 10 February 2022. Mr Martin told the inquiry the investigation was best left to the police and London Fire Brigade.

Ron Dobson, then commissioner for the London Fire Brigade, wrote to the government's chief fire and rescue advisor Sir Ken Knight to inform him of these results and warn him that the brigade had 'become aware that this type of panel has been supplied by more than one company'. 'In the circumstances, we believe it may be appropriate for a warning to be given to housing providers that it would be advisable to check the specification for external wall panels in their high-rise housing stock and check that what has been installed meets the correct specification,' he said.

But the government was not keen to take this step. Internally, one official wrote that they needed to be 'very careful' to avoid 'setting the hares running'. Another said such a warning would be 'critical information in housing terms' and 'we would need to assess what is to be made public'.[2] In the end, Sir Ken responded to Mr Dobson saying there was 'insufficient information' to warrant alerting housing providers to the findings and only a generic warning was sent, which told providers to check 'if there is any doubt over the compliance'.* Asked why the fire did not prompt an investigation of materials on other buildings, Brian Martin told the Grenfell Inquiry that it was 'not severe' enough to justify doing so.[3]

The upshot was that the critical information – that dangerous, non-compliant panels had ended up on a London

---

* Inquiry transcript, 1 March 2022. Sir Ken told the inquiry that he was concerned about prejudicing the police investigation if too much information was put into the public domain.

high-rise and had potentially contributed to the deaths of six people – was suppressed. Only the upper echelons of the London Fire Brigade and the government department responsible for building regulations knew. Some may have hoped that as more information emerged about Lakanal House, more concrete steps would be taken to prevent a repeat. But the political landscape was about to shift in a way which would make tightening regulations almost impossible.

'THIS COALITION HAS A CLEAR NEW YEAR'S RESOLUTION: TO KILL OFF THE HEALTH AND SAFETY CULTURE FOR GOOD'

On 6 May 2010, a general election saw the Conservatives become the largest party and the dominant force in UK politics once more. A coalition between them and the Liberal Democrats ended thirteen years of Labour Party rule and brought David Cameron into office as prime minister. And with a new government came a new ideology.

With the country, and the world, reeling from the global financial crisis, the government's answer was to radically shrink the state, in the hope that doing so would unleash economic growth. It embarked on a programme of austerity – cutting back and squeezing public budgets for every penny it could find. But the philosophy was broader. Under the banner of what Cameron called 'the Big Society', it set out a vision of private and voluntary organisations taking over responsibility for roles previously carried out by central government. 'Localism' would see power passed down to local decision makers and in what was branded a 'bonfire of the Quangos', state organisations were systematically closed down and their roles abolished. It was the most radical reshaping of the British state since Thatcher.

While all of these policy positions are relevant to the Grenfell Tower story, none had quite such a large impact as the final pillar of Cameron's efforts to cut back the limbs of government: deregulation. Under his leadership, regulation became a dirty word. In April 2011, a 'red tape challenge' was announced by the Cabinet Office. 'Excessive regulation is burdening businesses, hurting our economy and damaging our society,' said its press release.[4] From now on, the government would seek any and all ways it could find to reduce what it referred to as 'burdens on business'. Amid a belief that a recovery in new house-building – that had plummeted to historic lows during the recession which followed the banking crisis – was crucial to the economic recovery, the residential construction industry was placed at the heart of this drive. As part of the March 2011 Budget, the government adopted a rule which meant a special waiver had to be obtained to make any changes to the building regulations. An impact assessment warned that this risked breaching the state's duty to protect the right to life. In a chilling indication of its priorities, the government went ahead anyway.[5] In April 2011, Mr Cameron personally wrote to all ministers, with an instruction to 'sweep away unnecessary bureaucracy and complexity'. He said it was 'not a polite request to reduce regulation if you can' but 'a change in approach that means ministerial teams should see themselves person-ally accountable for the number of regulations contained within and coming out of departments, and the burden, they impose'.[6] In January 2012, he gave a speech. 'This coalition has a clear New Year's resolution: to kill off the health and safety culture for good,' he said. He pledged to 'wage war against the excessive health and safety culture that has become an albatross around the neck of British businesses'.[7]

This war took the form of a new policy for regulation: one in, two out. For every new regulation introduced, two had to be binned. This was to be measured financially: civil servants assessed the cost to business of any new rule and then had to cut enough restriction to remove twice that cost from the businesses affected. A former senior civil servant has told me this effectively made major new regu-lations impossible. Anthony Burd, who was responsible for Approved Document B alongside Mr Martin until 2012, described it as 'an effective moratorium' on introducing new rules. The policy would become 'one in, three out' in 2016.

All of this had a profound impact on the buildings regulation division, which sat within the Department for Communities and Local Government (DCLG), headed by Cameron-ally Eric Pickles. A champion of localism, Mr Pickles firmly believed in shrinking the influence of central government. He imposed this philosophy on his department. A former junior minister recalls that Mr

Pickles would decry 'regulatory madness' when new rules were suggested and branded officials '*Guardian*-reading pinkos'.[8]

'There was a deregulation focus that cut right across everything that was going on in the department,' one former official told me in 2019. 'It would have been very difficult for officials to suggest new regulations, because of what was going on in the department and the mood music.' Civil servants were strongly discouraged from offering advice ministers didn't like. The source describes it as a 'hostile environment'. Officials were reduced to tears if they presented reports to the ministers that went against the accepted consensus. 'The environment was toxic,' the source says. 'It was very, very difficult to give advice, there was a dislike of regulation generally.' In December 2011, Eric Pickles wrote to David Cameron setting out his plans for deregulation. He included the fire safety provisions of building regulations, promising to save businesses £25.4m per year by cutting rules.*

Richard Harrall, Brian Martin's line manager in the department, told the inquiry that building regulations were 'perennially marginalised' in the department, in favour of a focus on planning reform and a push to cut red tape. He said the appetite for deregulation was 'very tangible' in the department. 'The impact was absolutely evident, and discussions with ministers reinforced that,' he said.[9]

---

* Inquiry transcript, 7 April 2022. Mr Pickles described this as a 'mistake' insisting that fire safety rules were exempt from the deregulatory provisions.

Meanwhile, the government appears not to have wanted to hear news that it was wrong. In October 2012, the BRE's longstanding contract to investigate the implications of significant fires was renewed by the government. But the new contract contained a clause which said their reports should 'not contain any policy recommendations' or suggest amendments to official guidance. Such recommendations could only be made at the direct request of the department. 'This came after the general move towards deregulation, so regulation was not welcome,' one of the BRE's scientists told the inquiry.[10]

But this infatuation with cutting red tape was set to come into direct conflict with fire safety. Because in early 2013, the long-awaited inquest into the deaths at Lakanal House began. What would triumph? The need to prevent a repeat, or the desire to limit regulation at all costs?

APPROVED DOCUMENT B IS 'A MOST DIFFICULT DOCUMENT TO USE'

The inquest into the deaths at Lakanal House took fifty days. It exposed a series of major failures with the internal compartmentation of the building: fire doors failed, a suspended ceiling circumvented fire breaks meaning flames could rip through internal corridors, and ducts connected flats meaning smoke could pass to those where trapped residents were sheltering. Legally required risk assessments had not been carried out by Southwark Council, the landlord.

The coroner was sufficiently concerned about the official guidance on fire safety, contained in Approved Document B, to order the government to act. In a letter sent following the conclusion of evidence, she told Eric Pickles that it was 'a most difficult document to use'. She said it should be reviewed, to ensure that it was sufficient for refurbishments as well as new building work and was 'intelligible to the wide range of people and bodies engaged in the construction, maintenance and refurbishment' of buildings. Crucially, she also told Mr Pickles to ensure that it 'provides clear guidance... with particular regard to the spread of fire over the external envelope of the building'.[11]

Now was the moment to finally act on the terrible ambiguity which should have been obvious since fire ripped through Knowsley Heights in 1991. Class 0 needed to be ditched, the fudged 'filler material' fix adopted in 2006 abandoned and unambiguous guidance provided that combustible materials did not belong on the external walls of high-rise buildings. This was no longer a purely academic issue. Six people had died. More lives were at risk if the government did not act. Four years had already elapsed since Lakanal. It was time for change.

'WE ONLY HAVE A DUTY TO RESPOND TO THE CORONER, NOT KISS HER BACKSIDE'

It fell to Brian Martin to write a briefing note for ministers on the coroner's recommendations, and what could be done about them. The coroner had called for a straight-forward, focused look at whether guidance on the issue of external fire spread was appropriate. But Mr Martin suggested to ministers that she had recommended a full rewrite of Approved Document B. This, he said 'would require significant resources and have a disruptive effect on the construction industry'.[12] Instead, he suggested, the government could simply rework a certification scheme for window installers to ensure their members were aware of the rules – something the coroner had not recommended. A review of Approved Document B was already scheduled for 2016/17, and this could be drawn to the coroner's attention as well.

Martin had brushed aside the coroner's central concern about external fire spread. But ministers accepted his suggestion. Eric Pickles signed off on the response to the coroner, which told her the department was 'committed to a programme of simplification' and cited the unrequested review of rules for window installers. 'We have commissioned research which will feed into a review of this part of the building regulations,' he added. 'We expect this work to form the basis of a formal review leading to the publication of a new edition of the Approved Document in 2016/17.'[13] The changes had been kicked into the long grass.

The coroner did not only call for building regulations to change. She had also heard evidence about how a sprinkler system may have extinguished the fire before it spread from the flat where it started. As a result, she told the government to 'encourage providers of housing in high-rise residential buildings... to consider the retrofitting of sprinkler systems'.[14] But sprinkler systems were not popular with government. A report, written by Sir Ken Knight, said that while they could be effective in containing fire, their cost made them 'not practically or economically viable'.

In fact, the government had received similar advice from a coroner investigating the deaths of two firefighters in Shirley Towers, Southampton. A letter had been sent to housing providers as a result.

Officials did not see the need to send another one. Mr Martin called this: 'A big and essentially pointless task.' He said the department could 'tell the coroner that we've already raised this with social landlords… so we don't plan to do anything'. 'We only have a duty to respond to the coroner, not kiss her backside,' he wrote.*

Mr Pickles' letter attached a copy of the advice given to social landlords after the Shirley Towers inquest. Nothing more was done. Mr Martin later told the inquiry that the department was concerned that if it pushed councils too hard to fit sprinklers, it would be legally required to provide funding to pay for it. In an era of austerity, this was never going to happen.

But the aversion to sprinklers went further than this. In fact, three days after writing to the coroner Mr Pickles sent another letter to the Welsh Government accusing it

---

* Inquiry transcript, 29 March 2022. Mr Martin said this was an 'informal comment' and he was merely setting out the legal position that the department was required to consider the coroner's letter but not obliged to implement it.

of being 'over-zealous' and adding £13,000 to the cost of building new homes by requiring sprinklers in new builds.

Perhaps housing providers would act without being prompted by government? One housing management body for a local council considered the Lakanal House coroner's verdicts eighteen days before Mr Pickles sent his letter. It noted that implementing the recommendations, such as sprinklers, 'would have a significant impact on all landlords with responsibility for high-rise blocks'. However, the minutes go on to say that 'initial indications from [D]CLG are that these recommendations are unlikely to be taken up'.[15] As a result, nothing was done.

The name of this management body was the Kensington and Chelsea Tenant Management Organisation (the TMO). One of its properties was Grenfell Tower. Along with most social housing blocks around the country, it would not be fitted with sprinklers, which may have put out – for example – a routine fire in a kitchen fridge before it took hold on the exterior of the building.

'SHOULD A MAJOR FIRE TRAGEDY OCCUR IN A PURPOSE-BUILT BLOCK OF FLATS THEN THE GROUP WOULD BE BOUND TO BRING THIS TO OTHERS' ATTENTION'

While Mr Pickles had done nothing other than state the department's existing plan to publish a new version of Approved Document B within four years, this meant the small building regulations division was now committed to this timeline. If they were going to hit it, they needed to get going.

The first step was the publication of research reports on the guidance, which had been commissioned in 2012 and had to be published for public view before the government could start consulting on the changes to the Approved Document. Ministers simply needed to sign off these documents for publication. They were delivered to the department in February 2015. But the approval for their sign off was not forthcoming. Officials chased ministers' private offices for approval in May, July, August, October, November, December 2016, and January and March 2017. But they were simply focused on other things. In one email to ministers' advisers asking for approval, the research was described as 'very far from being a priority'. The research was ultimately not published until after the Grenfell Tower fire. The promised review never even got across the starting line.

What was happening in the meantime? The building regulations division found themselves tied up with work on a 'Housing Standards Review' – which was primarily deregulatory – and unable to make a start on the promised review of guidance.

In this period, Brian Martin received an even clearer warning that disaster was coming. In July 2014, he attended a meeting with industry experts hosted by the lobbying group the Centre for Window and Cladding Technology (CWCT). He left the meeting early, but received a copy of the minutes. Under the heading 'Use of ACM on high-rise buildings' it said 'the material generally achieves a reaction to fire classification of Class 0' and had been linked to a number of serious fires worldwide. They explained that the words 'filler material' was 'not clear' in banning use of the material. To clear up the confusion, they suggested an 'FAQ' be drafted by the BRE and published by Mr Martin to clarify that the deadly material was banned.[16]

But this never happened. The BRE never drafted an FAQ on this topic, and Mr Martin never published one. Instead, the change was pushed back to the full review of the Approved Document, which was stuck on the back-burner in the department.

Meanwhile, Mr Martin was well aware of the fires involving ACM around the world, but assured his colleagues it was banned in the UK. In May 2012, for example, following a blaze in France which tore up the outside of a tower block, he told colleagues: 'This [cladding] wouldn't be in accordance with Building Regs (probably).' In December 2012 fire tore through the 34-storey Tamweel Tower in Dubai, ripping from the ground to the roof in minutes. Asked by a colleague if he was aware of it, Mr Martin replied: 'Yes — have you seen the video? It's awesome.'

There was some external pressure to get moving. Sir David Amess, the Conservative MP for Southend West who would be murdered in his constituency office in 2021, was the longstanding chair of an All-Parrty Parliamentary Group convened to lobby for fire safety improvements, such as the installation of sprinklers. The group was advised by the campaigning architect Sam Webb, who had now helped the communities impacted by both the Lakanal

House fire and Ronan Point, and was acutely aware that the government was not doing enough to avert an even more severe catastrophe. Its secretary was Ronnie King, a former firefighter and a long-time advocate of sprinklers. The group wanted ministers to take account of new research carried out in 2012, which suggested sprinklers were now far more cost-effective, and abandon Class 0 in favour of a more restrictive fire performance standard for cladding panels.

They said this could be dealt with immediately by 'simple amendments' to official guidance, rather than waiting for the full review due in 2016/17. But these messages were not received well by ministers – who would frequently reply to lengthy letters with a brief two- or three-paragraph response.

On 9 September 2014, following several letters urging him to act, the minister then in charge of building regulations, Stephen Williams, wrote: 'I have neither seen nor heard anything that would suggest consideration of these specific potential changes is urgent and I am not willing to disrupt the work of this department by asking that these matters be brought forward.'[17]

Sir David responded to say he was 'at a loss to understand how you had concluded that credible and independent evidence which had life safety implications was not considered to be urgent'. He added: 'As a consequence, the group wishes to point out to you that should a major fire tragedy with loss of life occur between now and 2017 in, for example, a residential care facility or a purpose-built block of flats, where the matters raised here were found to be contributory to the outcome, then the group would be bound to bring this to others' attention.'[18]

The group never received a reply to this letter. Internally, officials were dismissive. 'Yes − he's very annoying,' wrote Brian Martin about Ronnie King in November 2014. 'He's miffed that we made some de-regulatory changes in 2013 so why can't we do a quick change to the [Approved Document] now and require sprinklers wherever they can go. Ronnie will not listen to reason, so I just ignore him.'[19] In an earlier email, he'd said putting Mr King in charge of Approved Document B would not 'necessarily be in the best interests of UK Plc.' Asked at the inquiry in 2022 what would happen if someone like Mr King was placed in charge of building regulations, Mr Martin bizarrely said he would 'bankrupt' the country and 'we would all starve to death'.[20]

In February 2015, there was yet another huge fire in Dubai, this time at the unfortunately named Torch tower. Once more, the building regulations division exchanged emails about the blaze. Brian Martin offered reassurance. 'There are provisions in the building regulations designed to prevent this kind of problem. So this shouldn't be a problem in the UK. There are, of course, no guarantees,' he wrote. He said nothing about the confusion around the word 'filler', or the call to clarify it which he had not acted upon. He would later accept that this reassurance lulled his colleagues into a false sense of security.[21]

With Mr Martin not preparing an FAQ to make it absolutely clear ACM was banned in this country, the chance to correct these fatal flaws was dependent on the delayed review of Approved Document B. In May 2015, politics changed once more. A general election removed the Liberal Democrats from office and brought a Conservative majority government to power. Jubilant, they doubled down on their deregulatory agenda. The review of fire safety guidance was now even less likely.

'WHEN IT GETS EXPOSED TO A FIRE, THE ALUMINIUM MELTS AWAY AND EXPOSES THE POLYETHYLENE CORE. WHOOSH!'

Following the 2015 election, officials decided to 'marry up' the review of Approved Document B with a broader review of all the Approved Documents. This was pragmatic. They knew they would be obliged by the 'one in, two out' rule to cut out costly 'burdens' on industry for anything they imposed. If they did the fire safety review separately, all the deregulation would be required from that same set of rules. Bundle it up with less life-critical matters – sound insulation for example – and there might be some wiggle room. But it also meant a delay to the process, which would not now complete until 2019 at best.

Meanwhile, the warnings of a looming disaster were about to get louder. In January 2016, there was another fire at a tower in the Middle East. In an email discussing it with his colleagues Mr Martin wrote: 'This stuff [ACM] is very rigid and makes nice shiny buildings. Sadly when it gets exposed to a fire, the aluminium melts away and exposes the polyethylene core. Whoosh!'[22]

Just a fortnight later, he would be contacted by a cladding supplier, Nick Jenkins, who had become concerned about the amount of ACM he was seeing sold in the UK. He'd recently told a conference that there could be 'an exact repeat of the Dubai fire in any number of buildings in London' and had emailed various industry contacts before being directed to Mr Martin. In the email, he told him that the number of buildings in the UK with ACM were 'many and growing' and most of them were 'installed in combination with various [combustible] foam thermal insulation boards'. He said the situation was of 'grave concern' and called on Mr Martin to provide 'a less ambiguous statement of the rules'.[23] This was a specific warning that the cladding which would ignite at Grenfell Tower eighteen months later was in use in the UK. The email was later described at the inquiry as a 'red alert situation' and 'plainly a life safety matter of the utmost importance'. Mr Martin knew the risks from the Middle East fires. He knew the guidance was confusing on this crucial point. He knew, now, from this email, that these materials were being used in the UK. Surely he would act?

He did not. Instead, he told Mr Jenkins that he was 'not sure the rules are all that ambiguous' and explained that he believed the words 'filler material' in the guidance 'could reasonably be considered' to ban the use of the panel. He added: 'If the designer and building control body choose to do something else then that's up to them.' Mr Martin told the inquiry he 'seriously underestimated' the hazard from ACM and the lack of deaths in Dubai had 'coloured my perception' of the hazard.[24] But the UK government

– via Mr Martin – had now effectively been told to expect a cladding disaster. It had chosen not to act.

In March, Mr Martin attended a follow-up meeting with the CWCT – the group which had asked him to provide an FAQ nearly two years previously. This time, minutes record that the group agreed that the guidance in Approved Document B was 'poorly written and open to interpretation' on this critical point. Mr Martin said it would be corrected in the next review of the document. All hope now hung on this review being completed before a blaze ripped through a building. Delaying it also allowed more and more blocks to be fitted with dangerous materials while government fiddled. But the review remained hopelessly delayed in a department refusing to prioritise it.

And there was still no progress from the ministers in charge. As the Tory majority government set about some of the most radical reforms of the social housing sector since Thatcher, working up plans to force local authorities to sell their highest value council homes and crank up the rents of social tenants on anything other than a minimal salary, building regulations were a forgotten footnote, a dusty policy area that received no attention. Mr Martin recalls that 'regulation was a dirty word' in the department. His senior, Richard Harrall, said the team was left working on a 'less than bare bones capacity'. 'I think that psychologically we were just beaten up, frankly,' he said.[25]

The All Party Parliamentary Group on Fire Safety continued to write to ministers, urging them to act. But they received a high-handed dismissal from the new minister responsible for building regulations, James Wharton, who cited the government's desire to 'reduce the burden of red tape'. Sir David Amess wrote to Mr Wharton to tell him that his comments had been met with 'unanimous dejection' from the group.

The government at this point justified itself by stating not enough people were dying in fires to give grounds for regulatory change. The number of fire deaths was coming down, so why impose more burdens on industry? But this was nothing to do with building regulations. Instead, smoking rates, reductions in chip pan use and better use of smoke alarms were all serving to drive down fire deaths. The figures were being misinterpreted as offering a false vote of confidence in a dangerously broken system of building regulations.

Sam Webb recalled that he sat next Mr Martin at a lunch in the Houses of Parliament in February 2016. He said he warned Mr Martin that another fire like Lakanal was 'likely' if the guidance was not removed and added that if it happened 'in the middle of the night when people were asleep, then the death toll was likely to be 10 to 12 times the six people who died in the Lakanal fire'. 'Brian Martin's reply to me was, "Where's the evidence? Show me the bodies",' said Mr Webb. 'It was as if he needed a disaster before he or the government would act.' Mr Martin denied using the phrase when asked directly by the inquiry.

Other sources have told me they have heard it as well. One was Arnold Tarling, a chartered surveyor and long-time advocate for fire safety who helped the APPG lobby for change after Lakanal. 'I said we have that product [ACM] in this country and it will lead to deaths as we don't have sprinklers,' Mr Tarling recalls. 'Nobody said I was wrong and everybody heard what I had to say, but nobody did anything. Their argument was that year-on-year the number of people dying in fires is falling and therefore buildings are safe. And then when you would say, "yes but we are not building like we used to, we are covering them with all these combustible materials that would go up like a horror movie", they would say "well it hasn't happened yet".' He recalls hearing the phrase 'show me the bodies' on more than one occasion.

The view that falling fire deaths meant no change was needed was frequently reiterated. 'The number of deaths in fires had been coming down very nicely. The [Cabinet Office] could not be convinced that there was sufficient danger to the public for major changes in regulation,' recalls David Sugden, former chair of the Passive Fire Protection Forum – which had issued several warnings about the need for change.

But, internally, even government figures knew this was a myth. In September 2016, a group of civil servants discussed the assertion that falling fire deaths proved the system of building regulations was working. 'Do we know whether the falls in fire deaths are ANYTHING to do with building fire standards, or are they caused by furniture standards changing, or falls in the number of people smoking at home/having open chip pans? Presumably if the number of deaths in older houses has gone down at the same rate, it is NOTHING to do with building safety standards, in which case the regs have achieved nothing,' wrote one senior advisor.

Mr Martin replied. 'As you suggest the reduction in the total number of reported fires is mostly attributable to a number of factors not related to building standards,' he wrote. These figures were only ever a smoke screen, a good soundbite to answer a difficult question.[26]

Mr Martin may also have been reassured by a report he commissioned from the BRE into the risk of external fire spread. This report was delivered in April 2016, with the conclusion that there was no major need for changes to regulations or guidance. 'With the exception of one or two unfortunate but rare cases, there is currently no evidence from these investigations to suggest that the current recommendations… are failing in their purpose,' it said. But the author admitted to the inquiry that the report was flawed. With an extremely limited budget, the testing was basic, incomplete and no survey had been carried out of the actual materials being installed on people's homes.[27]

And in the summer of 2016, British politics changed again. The country voted to leave the European Union, David Cameron left office to be replaced by Theresa May and the mantle of responsibility for building regulations passed to the new housing minister, Gavin Barwell.

'YOU CAN ROLL A TURD IN GLITTER IF IT GETS US MOVING'

When Mr Barwell was briefed by his officials on the building regulations issues he needed to deal with, they did not even mention the commitment to review Approved

Document B – such had its priority waned within the department.

Despite the coroner's deadline looming, Mr Barwell plunged his energy into a new Housing White Paper, which was supposed to provide watershed reforms to get more homes built. He focused on this exclusively through to spring 2017, with almost everything else relegated to the backburner.

Sir David's APPG kept on trying to warn him of the pressing danger – writing to request a meeting. But in October 2016, Mr Barwell declined to meet them and several of their follow-up letters got lost within the department.

In February 2017, with the White Paper finally out of the way, the building regulations team pulled together a series of documents to send to Mr Barwell to finally get the long, long delayed review of Approved Document B moving. They discussed how to couch the description of what had happened so far, in order to avoid looking like 'we've been sitting on our hands'. 'I'm pretty sure the department's upper echelons have sat on my hands for the last 18 months!' replied Mr Martin.

'They are not even aware your hands exist to sit on. Let's just agree to gild lilies where necessary,' replied Mr Harral.

'You can roll a turd in glitter if it gets us moving,' replied Mr Martin.[28]

On 2 May, Mr Barwell finally replied to the APPG, agreeing to meet them. But he brushed off their concerns about fire safety by citing the long-standing reliance on the stay put policy. 'Each flat is designed to prevent fire spreading to adjacent flats... experience of this approach to fire safety over many years has shown this to be an effective strategy,' he said.[29] Except, of course, at Lakanal House, which had by now been all but forgotten by the department.

Warnings about fire were also communicated to government by the LFB. Dany Cotton, then commissioner, wrote to Mr Barwell in April 2017 explaining her fears about the risks to high-rise buildings in London. She said the LFB had become aware of many blocks with 'significant compartmentation deficiencies'. The letter said these deficiencies had been discovered in at least one building every month since the start of 2017, and that 'it was safe to assume that there were many other cases that had not come to its [LFB's] attention'.[30] The letter never even reached Mr Barwell. Theresa May had called an election, and he was off campaigning in his marginal Croydon constituency – a battle he would ultimately lose.

But still, alerts were sounded. On the afternoon of 13 June, Mr Martin forwarded colleagues some research which had been carried out by the Fire Sector Federation. The front sheet warned that the approved document was 'out of date, placing businesses and communities all over the UK at potentially fatal risk'. 'I expect our new minister will have this shoved up his nose at some point soon,' wrote Mr Martin in the covering email.[31]

Nine hours after he sent this email, a fridge would catch fire in Grenfell Tower. Thirty years of deregulation now exacted its tragic, and ultimately avoidable, price.

'FOR THAT, I'M BITTERLY SORRY'

What to make of Mr Martin's actions in the eighteen years before Grenfell, when he failed again and again to correct the fatal flaw in Approved Document B?

It is tempting to heap the blame for what happened on his shoulders. When he appeared before the Grenfell Tower Inquiry in 2022, he was questioned about his actions, omissions and callously worded emails for seven days. At the very end, the barrister questioning him – the fiercely intelligent and forensic Richard Millett QC – presented a list of failings, one after the other, asking him to explain why he did not reveal them to his seniors in the immediate aftermath of the fire. At the end, Mr Martin was invited to accept 'they reveal a catalogue of failures, errors and omissions' and he responded: 'I think to some extent, yes.'[32]

He was then asked what he would have done differently, given his time again and knowing what he knows now. Finally, his evasive, irritable demeanour cracked. He said there were 'a number of occasions where I could have potentially prevented this [the fire] happening'. 'What I will say is that the approach the government – the successive governments had to regulation had had an impact on the way we worked, the resources that we had available, and perhaps the mindset that we'd adopted as a team, and myself in particular. I think, as a result of that, I ended up being the single point of failure in the department… For that, I'm bitterly sorry.'[33]

But was Brian Martin really a single point of failure? It was, in the end, the system the UK government set up which made him responsible for a policy area on which so many lives depended despite his woeful lack of qualifications. It was the UK government which adopted a position of deregulation and austerity. It was the UK government to whom the Lakanal House coroner addressed her recommendations, and to whom the select committee in 1999 delivered theirs.

While Mr Martin certainly seems to have shared some of the philosophy of his superiors, he did not decide to put them at the heart of policy making. Mr Martin, if Sam Webb's testimony is to be believed, may have been the one who actually used the words 'show me the bodies'. But the British state's consistent inaction for more than thirty years had said it all.

5

1.30 A.M.

Sakina Afrasehabi had never wanted to live in a high-rise building. Born in Iran, she had moved to the UK in 1997. Due to an accident years before, she suffered from serious pain and mobility problems, as well as arthritis and other medical conditions. Now aged sixty-five, she found it extremely difficult to navigate stairs.

She had first been housed by the local council in a flat in Ladbroke Grove, not far from Grenfell Tower. But this flat was inappropriate for her: there were forty-two steps to the front door and no lift. Her medical condition rendered her practically housebound without assistance.

Because of this, her family had applied for transfer to a suitable property in 2000. By 2012, with no prospect of a move and twelve years on the waiting list for a new home, her family instructed a solicitor to bring a case against the council. She was told she would have to wait. Some families, the council said, had spent twenty years on the waiting list.

The council had ruled that she should not live in a property with more than six steps, or above the fourth floor in a lifted building, and in 2014 she found a ground floor home she liked and prepared to move. But the family were suddenly put under investigation by the Royal Borough of Kensington and Chelsea's (RBKC) local council, who raided the flat, ransacked wardrobes and called them in for questioning at the town hall. They suspected, without evidence, that her daughter Nazanin was not really living with her. No wrongdoing was found, but Sakina lost the chance to move into the flat as a result.

In early 2016, a flat became available in Grenfell Tower on the eighteenth floor. When they viewed it, the property was a mess. Doors were hanging off cabinets in the kitchen, gas pipes were exposed and workmen were busily and noisily at work refurbishing the tower. The family were worried about how their mum would cope so high up, especially if the lifts broke. They were told they had to take the property or be suspended from the system.

When she first moved in, the refurbishment meant there was no lift access from the ground floor, and she had to walk up two flights to reach the lift. 'I had deep worries at the time but I buried them in the back of my mind. I did not want to even contemplate what would happen if a fire ever broke out,' Nazanin recalled later.

Living in Grenfell Tower was difficult for those with disabilities. If the lifts were out of service (and they often were – particularly during the refurbishment when one was given over to workmen), they were trapped in the building. The tower had a narrow, single staircase located next to the lifts on each landing. But it had no adaptations to help those with mobility issues navigate it. And no plans had been made for how people with disabilities could leave the tower in an emergency. As we will see – this was true of almost all high-rise buildings at the time, thanks to government-backed guidance that said the provision of escape plans was unnecessary. Despite these troubles, Sakina lived a happy life. She had a strong sense of humour, warmth and was an amazing cook; her fish stew was famous in her family. She was very close to her family: her five children and her sister Fatemah. Fatemah, six years younger than Sakina at fifty-nine, was known for her artistic talents, the dolls she made and her beautiful singing voice. The two women had worked to prepare a large meal for the family on 13 June, including some of Sakina's children and grandchildren. It was a happy evening. The rest of the family left at around 10 p.m., but Fatemah stayed the night.

At around 1.30 a.m., the sisters were warned of the fire by a neighbour who banged on their door. They left their flat on the eighteenth floor and entered the stairwell. But they did not go down. Instead, along with a group of fifteen other people, they headed upwards towards the top floor.

Why this happened is not entirely clear. Certainly, there is some evidence that they were told to go back by one of the many firefighters in the stairs at this point. One resident, who escaped at around the same time, recalls hearing 'shouting in a clear English accent, "go back, go back"'. This was said to have the effect of 'causing people to turn around in panic'. Other residents who escaped were told to go back to their flats by firefighters they met on the stairs, but ignored them. It is also known that an attempted rescue a couple of floors below resulted in a door being held open which would have caused smoke to billow suddenly into the stairs and create a perception of immediate danger. Some residents appear to have believed a helicopter would rescue them from the roof.

For some though, the simple difficulty of descending without help would have been a barrier. A neighbour met Sakina on the landing and told her the lifts were not working and she needed to use the stairs. But she had pointed at her legs to indicate she could not. Fatemah would later say on a phone call with her nephew that they had been 'told' to go up, but she did not say by whom.

Another who went up the stairs was Debbie Lamprell, who lived on the nineteenth floor. Debbie had lived in west London since she was a young woman. She loved her flat. She had many friends in the tower and loved her job at the nearby Holland Park Opera House.

Debbie's mother Miriam would visit her regularly in the flat and neighbours would comment on the noise the two women would make laughing together. In recent years though visiting had become more difficult. The lifts were frequently out of service and climbing the stairs to the nineteenth floor was difficult for Miriam. Nonetheless, Debbie would still text her mother every morning and phone her every evening, or in the afternoon if she was working a late shift. Before going to bed on the night of 13 June, she had texted her mother to say 'Hi mum, I am home alright, good night and god bless, lots of love.' She texted her mother every night to let her know she was safe.

But now, at around 1.30 a.m., she was awake and afraid – there was fire in the block and smoke in the lobby. One of her neighbours recalls meeting her on the landing of the nineteenth floor where they lived, when the smoke was 'dark grey and steamy and very hard to see through'. 'She looked frightened and said to me that people were going upstairs. I thought that she had instructions to go upstairs,' the neighbour recalls. Debbie, her friend Gary Maunders who had been visiting her, and the neighbour all went upstairs. 'It was clear that fire was lower down the tower and walking down towards the fire made less sense,' the neighbour recalled later.

And so, fifteen people began arriving on the very top floor and knocking on the doors of their neighbours asking for help. The smoke was even worse on the top floor than it had been on lower floors. It was 'very dark and very thick' and 'smelt of chemicals,' said one. There is some evidence that the tower's smoke extraction system, which should have pulled smoke out of the landings lower in the tower, was leaking: smoke from the lower floors was seeping out into the landing at the very top.

Debbie and a large group of others entered Flat 201. Sakina, Fatima, and a 64-year-old mother and her 27-year-old daughter who had fled upwards from the twenty-second floor went into the home of a family of three: two parents and their grown-up son, in Flat 205. Others were greeted by Rania Ibrahim in Flat 203.

Two others – an elderly woman and her grown-up son – went to Flat 202. That flat was rented by a young Italian couple, Gloria Trevisan and Marco Gottardi. They had met at the Venice University Institute of Architecture, from which they had graduated in 2016.

After struggling to find satisfying work in Italy, they had moved to London in March 2017 and rented a flat on the top floor of Grenfell Tower. Gloria missed the sunshine and the food of Italy, but she liked living in London. In particular, she was pleased to have quickly found work at the architectural firm Peregrine Bryant – which specialised in the restoration of old buildings – her dream job. She was twenty-six, it was the first time she had been away from home for a prolonged period and she would speak to her parents on the phone almost every day.

Gloria would tell her mother about the incredible views she could see over London from the flat. She sent photos of the rainbows arcing over the city to her parents. Marco too had been lucky. He had found work at a studio called CIAO (Creative Ideas and Architectural Office). The couple liked the flat at the top of Grenfell Tower and the panoramic views across London.

On 13 June, they were at home together. They had considered returning to Italy for Gloria's parents' thirty-seventh wedding anniversary on 14 June, but had instead booked tickets to fly back for Marco's birthday in eight days' time.

At 1.34 a.m., Gloria phoned her mother in Italy and wished her a happy wedding anniversary. Then she told her that a fire had broken out in the building. While she was on the phone, the couple welcomed their neighbours from a few floors below into their flat.

Gloria's mother told her to cover her mouth with a cloth and flee the building, but the couple declined. Marco told her the fire brigade was there and they had been told to stay where they were. Also, they could see the landing filling up with thick, dense smoke and Gloria told her mother she did not think they would be able to escape without help.

*

Most of the cladding fires seen around the world before Grenfell went up the building in a straight line, stopped and burned themselves out. They did not wrap the building in an all-engulfing fire. Why did this happen at Grenfell?

At 1.30 a.m., fire extended up the building in a vertical, rectangular column – reaching straight up from Behailu's kitchen to the roof. The windows and kitchens in this column were now on fire, but the rest of the building was unaffected.

However, when the fire reached the top of the building, it ignited a feature known as the 'architectural crown'. This was added in 2012 because the council's planning department said the original designs looked 'dull' and something was necessary to 'accentuate' the top of the building.[1] As a result, a 'crown' was designed by the architects Studio E – a series of wing-shaped, curved panels fitted to the top of the building. They were comprised entirely of ACM, like the cladding. This feature had no purpose beyond aesthetics. And so, on the night of the fire, as the flames licked up to the top of the building, it ignited and acted – in the words of one expert witness – as a 'fuse', which took the fire steadily around the top of the building in both directions.[2] All those who had fled upwards were now imperilled.

*

At 1.30 a.m., a fire service command unit – codenamed CU8 – arrived at the base of the tower. These are mobile command centres, equipped with white boards and computers, and provide a base for fire service officers to command a major incident.

The van had a phone which could be used to communicate with the control room in Stratford. It was also equipped with an IT system which can bring up maps and pictures and record plans. It is also supposed to be able to connect to the video images being broadcast by the police helicopter and give those on the ground an overview of calls coming into the control room. But on the night of the fire, this technology was not working. This was not unusual. One firefighter told the inquiry in a witness statement that the command unit's IT system was notoriously unreliable. They ran on a Windows XP operating system – which was released in 2001 and was several generations out of date by the time Grenfell happened – and use a 3G signal to connect to the internet which is patchy and unreliable. The firefighter said he had 'sent hundreds' of reports about the system not working, but with 'nothing ever changing we all just stopped doing it'.

Without functioning IT, the initial system to record where residents were trapped in the tower was rudimentary. Calls would be phoned in to the officers aboard CU8. They would then radio through to an officer standing outside the tower who would scribble it onto a blank piece of A4 paper which would be passed along to the bridgehead inside the tower. A crew with breathing apparatus would be dispatched to find the trapped residents.

But getting up and around the tower was becoming increasingly difficult. Smoke filled the lobbies of the tower on at least eight floors. Residents who opened their doors on these floors were discouraged from leaving by the smoke, which tasted toxic and made them gag.

One such resident was Nick Burton, who lived on the nineteenth floor. He lived with his wife Pily and a beagle named Lewis Hamilton II. By 2017, the flat had been Pily's home for forty-two years. She was one of the first residents to move in after the tower was built in 1974.

Nick had grown up nearby. The area was formed as 'a community which was built up of layer upon layer of people coming into the area; people who came into this country to help build things here,' he recalls. His father was a bus driver from St Lucia. Pily's family came from Galicia in Spain. Her family had moved to London when she was a teenager and bought a large house in Kensington for £6,000. Pily moved into Grenfell Tower when she left her family home in her twenties.

Nick met Pily when he was in sixth form. She was flamboyant and outgoing and she loved to dance. Nick thought she was out of his league. After the discos, they would go back to her flat in Grenfell Tower with a group of friends and play cards and listen to music. Slowly they fell in love, and Nick moved himself in 'sock by sock'. They bought the flat through the Right to Buy scheme in 1994.

In 2000, they finally married, having been a couple for sixteen years. Popular among the Lancaster West community and known throughout west London's Spanish community, Nick and Pily's flat was open to everyone. It was always full of Galician music which Pily and her parents would sing as well as reggae which they all loved. Everybody who came by the house wanted to eat Pily's paella and Nick, who worked in the wine trade before becoming a hospital catering manager, always ensured there was always a steady supply of drinks. A lover of fashion, Pily was well known around the nearby Portobello Road market for her colourful and stylish outfits.

But life in the tower was made difficult by the poor maintenance and the often broken down lifts. 'There was no care. We understood that something was going to happen and maybe we were going to lose the estate,' Nick says.

After Pily's mother, father and brother died in quick succession in the 2010s, Nick took her away travelling around Europe to help her recover from the grief. They drove around the French wine regions, through Switzerland and to Milan, for Pily to enjoy the fashion. Then they drove to Venice, drank champagne on a gondola, had five days on the Italian beaches and then three in Rome. But something wasn't right.

When they returned home, they discovered Pily was beginning to suffer from dementia. Her condition worsened and she was forced to leave her job. For such an extrovert, this isolation was crushing. But Nick continued to support Pily and the couple got on with life, adjusting to the reality of her new illness.

On the night of the fire, Nick had been out to walk Lewis Hamilton II with some friends from the low-rise blocks next to the tower. As he returned home, he saw some vehicles parked in front of the entrance to the tower. 'Look at those idiots over there parked by the main door. If there was a fire in the tower, how would the fire truck get in?' he commented to his friend.

Thinking little more of it, he returned to his flat and put on a DVD with Pily. They fell asleep in the living room, Pily on the sofa and Lewis Hamilton II lying beside Nick on the floor.

At 1.34 a.m., he heard noises outside his front door and opened it. Black smoke came rushing into the apartment, the smell reminiscent of burning tyres. He could not see the opposite wall, no more than a metre and a half away. He shut the door and placed wet towels at the bottom.

He was aware of the stay put policy and worried that Pily, in her frail condition, would struggle to survive the smoke. He helped her dress, and they stayed in the flat together, waiting for help.

*

Grenfell became the deadliest cladding fire the world has ever seen not just because of the speed with which flames ripped up the outside of the building, but because of how fast smoke spread through the building. This shouldn't have happened. The tower should have been built to withstand the rapid, internal spread of smoke and flame.

The critical failure is believed to have been the entrance doors to the individual flats. As we will see in later chapters, they were defective: incapable of withstanding smoke and flame for even half the time required by official standards.

But on many floors, there was a bigger problem: they were open.

Fire doors are supposed to swing shut by themselves. They are fitted with devices called self-closers to ensure this happens. But at Grenfell there were endemic problems with these devices. Investigations after the fire showed forty-three front doors in the tower had no self-closers installed and thirty-four were not working. This meant that two-thirds of the doors from the fourth floor up would not swing shut by themselves when the residents left.

When people run for their lives, they don't stop to shut the door. This is simple human nature. And so, as the flames reached the flats numbered '6' on each floor, broke the windows and started fires in the kitchens, black smoke came rolling in from burning plastic on the walls of the tower. The fire then began to devour the contents of the flat – appliances, furniture, clothes – all of which let off toxic smoke when burnt. The residents fled, the door stayed open and the smoke poured out onto the landing. Now anyone who remained in the other flats on that floor would need to escape through the smoke to get out.

Many who opened their doors to a wall of black smoke simply slammed the door shut and phoned 999. One recalls: 'I felt like I'd been hit by gas as well as smoke, so basically it would stop me breathing.' The thick, toxic smoke on his landing forced him to rush back inside and rinse his streaming eyes.

Smoke had also begun to enter the stairway. Firefighters had propped open some of the heavy fire doors separating the lobbies and the stairs to run their hoses through. Smoke was passing through these open doors into the stairwell and steadily filling it up. And each time someone made it from a smoky lobby to the stairwell, they opened a door and smoke would follow them. It was, at this stage, still possible for those who could descend the stairs to escape alive. But, bit by bit, the stairwell was getting less passable.

*

Eddie Daffarn was born and raised in west London. He loved the area. But his home environment was complex and as a result he was sent to Woolverstone Hall School, near Ipswich – a former boarding school that had been obtained by the London County Council to educate boys from difficult backgrounds away from home. 'I learned how to learn, which has been useful,' says Eddie. 'But I didn't learn much else.'

When he finished school he returned to London and started working odd jobs in construction and painting and decorating. 'My mum described me as a bohemian, which is a nice way of saying I was a bit of a waster,' he recalls.

By the late 1990s, with Eddie approaching forty, his father died which he says served as a 'wake-up call'. He went into adult education, got a degree from Brunel University and became a mental health social worker. He'd picked up a sense of social justice from his grandad, who was at one time PA to the MP William Wedgwood Benn. 'I've only voted once in my life, I'm not a member of any political party, but I did inherit a bent towards social justice,' he says.

His time as a social worker in west London instilled in him a drive to organise and campaign. And there was plenty to campaign about. In the early 2010s, with austerity cuts biting, he was seeing things close down or be threatened with closure – the local library and adult education centre – while the council continued to pour hundreds of thousands into the local opera. 'They saw themselves as a property developer masquerading as a council,' he says. 'We just felt it really acutely, this library had been in the community for 125 years. People would tell you stories about how it had been important to them and their parents and their grandparents. In North Kensington, you're never more than one person away from someone who hasn't benefited from the college. The idea that it was going to be closed and redeveloped for housing came from someone who didn't understand that community full stop.'

While he loved the community in Grenfell Tower, the maintenance of the estate was woeful. 'There was definitely a feeling that our housing was being run down and that was not a good feeling,' he says.

But these problems helped build the community. 'The lifts were so often broken down that we came to know each other from standing in the lobby complaining about the fact that the lifts weren't working,' he says.

In the early 2010s, his activism came home. A series of local concerns, including plans to build a school and leisure centre outside the tower, angered the community. Eddie was right at the heart of this movement – an organiser and a mouthpiece for other residents who did not have the confidence or time to push through the TMO's exhausting complaints process. The refurbishment work, which we will cover in Chapter 10, further united the tower's community. 'We were coming together with a feeling that we weren't being treated right,' he says. 'These were our homes. It felt like an assault on our homes and people felt quite protective.'

Alongside a neighbour on the estate, Francis O'Connor, Eddie started a blog – the Grenfell Action Group – which chronicled his mounting dissatisfaction with conditions in the tower. In particular, Eddie was worried about fire. That summer there had been a worrying fire at another tower block in Shepherd's Bush, only a mile from Grenfell Tower, which had spread over five floors. He decided to commit his worries to his blog. In a post titled 'Playing with Fire!', he wrote: 'It is a truly terrifying thought but the Grenfell Action Group firmly believe that only a catastrophic event will expose the ineptitude and incompetence of our landlord, the KCTMO, and bring an end to the dangerous living conditions and neglect of health and safety legislation that they inflict upon their tenants and leaseholders.

'It is our conviction that a serious fire in a tower block or similar high density residential property is the most likely reason that those who wield power at the KCTMO will be found out and brought to justice!'

On the evening of 13 June 2017, he was alone in his flat on the thirteenth floor. It looked out west over the Tube lines and had beautiful views of the sun setting as London gave way to the beginnings of countryside and Heathrow airport. He turned on a Radio London phone-in talk show and went to bed.

He was woken up some time after 1 a.m. when he heard his neighbour's smoke alarm beeping and people shouting in the lobby. When he opened the door a gush of what he called 'thick, swirling acrid smoke' came rushing into his flat. He quickly shut the door again. He had seen that the whole lobby was full of smoke. Just then, his mobile started ringing and a friend outside the tower told him to get out. He went to the bathroom, wet a towel and put it over his head. Pausing to grab his mobile phone and keys, he stepped out into the lobby, closing his door behind him. 'The smoke was so thick I couldn't see beyond the end of my nose,' he said. He held his hand out trying to find his way to the stairs, but could see nothing.

'I was now in sheer panic,' he recalled. 'I started to use both hands to try and find the way out. I was running my hands along the wall but not finding the door. I don't recall dropping the towel away from my face but I think I may have done in my panic to find the way out. I started to inhale the smoke. I thought to myself "shit, man, this isn't going to end well for me". I thought I was about to die.'

Suddenly he felt a tapping on his leg and looked down to see a firefighter lying on the floor where the smoke was much thinner. He got down, and realised the firefighter was lying half in and half out of the door to the stairs. Seeing the way out, he ran, crouched beneath the smoke. In the stairwell, the smoke was much clearer – at this point, the large lobby doors were keeping the smoke out. Eddie flew down the stairs, running for his life. CCTV footage shows him exiting the building at 1.35 a.m.

Outside, he looked up at the tower and saw the flames ripping up the outside. 'I could see people in the windows who were still in the building,' he recalls. 'I was wailing from inside my soul. It was just so horrible. I don't know what floor they were on or who they were. After about five or ten minutes we were progressively being moved further and further back away from the building. I was in distress. I remember a Muslim man, a stranger, offering me comfort, putting his arm around me and asking if I wanted to go to his house. I could not stop crying.'

*

In the 10 minutes from 1.30 a.m. to 1.40 a.m., thirty-six people left the building safely having made it to the stairs.

'It was a little bit foggy. Like a foggy day. It was smoky but not the horrible dense black smoke like the lobby,' said one resident of the fifteenth floor in his witness statement. On his way down he met a neighbour who had polio and walked with crutches. He met him on the stairs near the ninth floor, as he was clinging to the handrail and trying to drag himself down the stairs. The resident from the fifteenth floor stopped. 'We are going to go down together,' he said in Arabic.

'He was panicking. I was panicking too. I put my arms under his arms. It was quite difficult for him. As we went down we were parallel. He was still holding the handrail but I had to let go of it and there was no handrail on the left hand side for me to hold,' he recalled in his witness statement.

'We then went slowly downstairs together. I was struggling to breathe; it was really bad. The smoke was getting worse. He was heavy. It was so very slow. There was no way for me to go quickly and no way for him to go quickly. He helped me though; he gave me a purpose.'

Both men made it out from the tower alive.

At 1.38 a.m. Rania Ibrahim was standing by the front door to her twenty-third-floor flat, a smoke alarm blaring. She was pondering whether or not to open her front door because she thought there was someone on the landing, which was becoming increasingly full of smoke. She took out her phone and began to stream live to Facebook. A voice behind her told her not to open the door. There was audible banging on the landing.

After pausing for a moment, and praying in Arabic, she opened it and called to the people outside to come into her apartment. 'Hello, come here,' she shouted into the darkness. 'Come here, hello.'

'I'm here, here,' a cockney voice replied, and a white man, who had fled from a lower floor came into the apartment. There were more voices on the landing but it was pitch-dark with smoke. Rania continued to call into the darkness.

'The smoke is coming,' said a man behind her, and reached over to shut the door. By this point, Rania had invited a number of people who had come up from lower floors into her home. Her two daughters, Fethia and Hania, were also inside the flat.

Still worried that there were people trapped on the landing, Rania stepped out into the smoke. She continued to call 'hello' and a voice replied. It was Marco Gottardi. 'We are here. I am inside our apartment,' he said.

Relieved that no one was trapped, Rania closed the door. She went to film out of the window. Blue flashing lights rapidly approaching the tower were visible. 'There is no God but Allah,' she said in Arabic, coughing slightly as she spoke.

'We're stuck on the twenty-third floor,' a voice screamed from a neighbouring window.

Outside the tower, her sister in the UK had driven from her home nearby and arrived. She fell to her knees when she saw the flames and a press photographer took a photo of her silhouetted against the blaze. It would appear on the front pages of national newspapers in the days to come.

In the control room in Stratford, bewilderment prevailed. They were being inundated with calls from residents saying they were being affected by the fire. They simply did not know how this was possible: the fire was on the lower floors, so how could people all the way up the building have smoke and flame outside their windows? The number of fire engines attending was now twenty-five. A senior manager had been paged.

'Please can you hurry up?' a young girl on the top floor begged the operator she was speaking to. Smoke was starting to fill up the flat. She was advised to block up gaps in the doors and wait for rescue. 'Please hurry up,' she said. 'I can't breathe.'

'VERY CONFIDENTIAL!'

It is time to explain how a cladding panel with a core of solid petrol found itself on the market for use on the walls of people's homes.

Aluminium Composite Material (ACM) was first invented in the 1960s. A German company realised that bonding the metal to plastic made it easier to work and cheaper to produce, and it was initially used in furniture. From the 1990s onwards, it became a popular choice for building facades: providing the elegant curves and crisp finishes modern designers craved.[1]

There are several large manufacturers of ACM in operation. One would ultimately sell the cladding panels on Grenfell: Arconic. This is a company with deep roots, having been spun out of Alcoa, which traces its own history back to the Aluminium Company of America – a giant beast of the American materials sector which was founded in 1886 when the inventor Charles Martin Hall discovered the process of smelting. Alcoa provided the metal for the spacecraft used in the first moon landings, US military hardware from the battlefields of World War I to the bomber jets which rained death over the Middle

East in the early 2000s and drinks cans which litter the environment all over the globe.[2]

The panels were sold for use on Grenfell by the European arm of this company, AAP SAS, which is based in north-eastern France. The fire safety risk of these panels was discovered in 2004. They could have been taken off the market then. But no one then was willing to risk corporate profits for basic fire safety.

A DEADLY SECRET

There are two primary ways of fixing ACM panels to the wall of a building. One is to drill holes in them and bolt them to a frame with rivets. The other is to bend them into an 'L' shape so they can be hung on hidden rails. The latter, known as 'cassettes', is slightly more expensive, but is seen as neater: the rivets disrupt the smooth facade of the building and can become rusty.

In 2004, with the same new alphabetical Europe-wide testing standard which would trigger warnings about the need for revision of UK standards about to come into force, Arconic's French arm commissioned tests on its cladding panels, both fixed to a wall with rivets and bent into cassettes.

While the riveted panel gained a 'B' grade – making it suitable for high-rise buildings in many European countries – the cassette system was a disaster. It burned ten times as fast, released seven times as much heat and three times as much smoke. The test had to be stopped early so the panel could not even obtain the basement 'E' ranking.

This was a catastrophic failure and a clear warning that this material in this shape was horrendously unsafe.

But Arconic did not stop selling it and did not tell the markets about the testing. Instead, its technical manager Claude Wehrle dismissed the test as a 'rogue result' and issued no warning to its clients about what had happened. Claude Schmidt, the president of the company, would later deny that this test was the firm's 'deadly secret'.[3]

A MISLEADING HALF TRUTH

ACM comes in three main varieties. One has a core of pure polyethylene, making it the most flammable. The second category has a core where the polyethylene is diluted with minerals, making it less dangerous in a fire. And the third has an entirely plastic-free core, making it non-combustible.

Arconic, following its 2004 test, could have pulled the combustible version out of the market and only sold the less dangerous panels. In several jurisdictions, which adopted higher standards, this is essentially what it was required to do. Countries such as Germany became considered 'FR' markets – standing for 'Fire Retardant' – because they set tough standards for how flammable buildings could be.

But this was not the case in the UK. The government's failure to tighten the Class 0 rating meant the cheaper panels could stay on the market. The pure polyethylene was around £2 to £3 cheaper per square metre, which gave it an edge when construction firms tried to squeeze every penny out of their costs through a process known as 'value engineering'.

All Arconic needed to do was persuade builders that their polyethylene-cored product complied with the requirements of Approved Document B. To do this, they sought external certification from the UK construction industry's most trusted source. The British Board of Agrément (BBA) was initially established by the government in the 1960s to certify construction products as suitable for use. It became a private organisation in 1983 – reliant on commercial income to fund its work.

Its business model is to charge product manufacturers a fee to assess their products and then offer a certificate which confirms the claims in their advertising. On its website it advertised itself as 'the leading authority on building product certification' and called its certification the 'vital ingredient in the provision of assurance, quality and integrity'.[4] For product manufacturers, a BBA certificate was a passport to winning sales.

Arconic approached the BBA seeking a certificate for its ACM in November 2006. It agreed a fee of £16,500. According to its contract, Arconic was supposed to provide all of its relevant test data to the BBA. But it omitted the information about the 'cassette system' two years earlier. Instead, it only shared the test in which the panel had been fitted to the walls with rivets. Despite this, the certificate contained a diagram showing both fixings. Claude Schmidt would later accept this was a 'misleading half-truth'.[5] In its closing statement, Arconic insisted the certificate was not misleading – saying that it applied to the flat sheet, which is what the company sold, making its fire performance once fabricated into cassettes irrelevant.

They also argued Mr Schmidt's answer was based on issues with the translation of the question.*

It also provided tests claiming to meet the Class 0 standard. But it only provided a 2003 test on a 'fire retardant' version of the product. No Class 0 test was supplied for the polyethylene version. Nonetheless, the BBA happily published a certificate asserting that the product – in both forms – 'may be regarded as having a Class 0 surface'. This would be enough to persuade many contractors that it was safe and compliant for use on high-rise buildings. Arconic's UK salespeople began to win job after job for new builds and refurbishment projects around the country: 45,500 square metres in 2007, 69,482 square metres in 2008.[6] The company pulled in €3.5m from these sales in two years. Job by job, the walls of our tower blocks were fitted with violently combustible plastic.

'WE ARE NOT "CLEAN"'

In September 2007, Arconic's marketing manager Gerard Sonntag took a corporate trip to an industry conference

---

* The firm argued that rather than the French word '*trompeur*', which carries the English meaning of deceptive, Mr Schmidt answered on the basis of the understanding that the question meant 'wrong such as to lead a person into error', even inadvertently. In its closing statement, Arconic said that he was doing 'no more than stating the obvious' but had not accepted the company acted deliberately to mislead.

in Norway. According to Mr Sonntag's note of the trip, one speaker raised significant concerns about the use of ACM. He said the use of 5,000 square metres of polyethylene-cored ACM on a building would have the same 'fuel power' as attaching a 19,000-litre oil truck to its walls. 'Let's imagine that Otefal [the speaker's company] organise a lobbying activity on the European [Parliament] and show such a presentation... the result could become catastrophic for the ACM products,' he wrote in a note about the trip. 'One of the arguments from Mr Pohl was, "what will happen if only one building made out PE [polyethylene] core is in fire and will kill 60 to 70 persons, what is the responsibility of the ACM supplier?"'[7]

This is a devastating piece of evidence – revealing that the political consequences of a disaster like Grenfell were actively under consideration within Arconic ten years before the fire. And yet, the product stayed on the market. Mr Sonntag suggested a project to lower the cost of FR panels to the same as PE. This was taken up, but PE remained the default choice in the UK.

Meanwhile, in France, technical manager Mr Wehrle was becoming increasingly concerned about the risk of a big ACM fire. 'Here are some pictures to show you how dangerous PE can be when it comes to architecture,' he wrote in an internal email in 2009 after a fire at an office block in Bucharest.

When a client in Spain emailed to ask about the fire rating of the panels in 2010, he explained to Arconic's salesperson that they did not perform as well when bent into 'cassettes'. But he did not suggest telling the client this. 'This shortfall in relation to this standard is some-thing we have to keep VERY CONFIDENTIAL!!!!' he wrote. Asked specifically to provide a certificate for the cassettes, he sent another internal email saying: 'It's hard to make a note about this… Because we are not "clean."' He would later refuse to attend the public inquiry in the UK to answer questions about these emails.[8]

In 2011, further tests were finally commissioned on the cassette panel – but they performed in the same way as those tested in 2004, burning ferociously. In October 2011, they were formally categorised as an 'E', meaning they were inappropriate for high-rises in almost all of Europe. By now Mr Wehrle had a theory. The bend in the panel to hang it on the rail meant the polyethylene col-lected as it melted, heated and ignited causing the entire panel to burst into an uncontrollable fire. But still Arconic kept selling it. In 2012 – following an ACM cladding fire in France that killed a disabled woman – the company removed the 'B' grade from its marketing material. But it did not announce that its correct grading – at least in cassette form – was an 'E'.

The company was now increasingly resigned to having to sell the more expensive fire-retardant product across most of Europe. But luckily for them, there were still some countries where regulations lagged behind. 'For the moment, even if we know that PE material in cassette has a bad behaviour exposed to fire, we can still work with national regulations who are not as restrictive,' minutes of a November 2011 meeting attended by Mr Wehrle say. One of these markets was the UK.

'A CHIMNEY WHICH TRANSPORTS FIRE'

Arconic's salesperson in the UK was Debbie French. She knew little about fire regulations and believed the Class 0 rating was what mattered in winning jobs. She would send the BBA certificate out to clients to confirm the suitability of the product and ask few further questions.

But she did have some awareness that the material could be dangerous. In April 2013, following one of huge cladding fires in the UAE, she was forwarded an email from a competitor. This firm was telling its clients its 'FR' product was safer than some of the others on the market.

Its salesperson warned an ACM cladding system could act as a 'chimney which transports the fire from top to bottom'. He said that when informal tests were shown to architects in Bangkok 'they nearly fainted'. This email was passed to Ms French, and she forwarded it to her bosses in France, asking what she should do.[9]

Ultimately, she decided to address the issue head on and offer her clients reassurance. Linking to a BBC report about one of the fires in the Middle East, she wrote: 'As a business we are aware of this report and our technical team are following the details... Regarding the supply of Reynobond in the UK, as you know we supply both PE and FR core and can control and understand the core that is being used in all projects. At this stage we will continue to offer both FR & PE and continue the close working relationship we have with our approved fabricators to make sure the right... materials are being used and installed on all Reynobond projects.'

Asked about this email later, she would disavow it – insisting it was 'too heavy on the sales side' and did not reflect the true position. Claude Schmidt would also reject it – saying such specific control was 'never our policy'.[10]

In November 2013, further tests once more confirmed the 'E' grade of the cassette panels and this time bumped even the riveted system down from a 'B' to 'C'. Arconic finally decided to make the results clear. It classified both panels as an 'E' and wrote to its salespeople across Europe to inform them of this change and tell them the 'B' grade should no longer be relied on. This email went to Debbie French on 3 February 2014. But by this point, she was already lining up a big sale of ACM to a tower block in west London and she did not tell clients about the change in the rating of the product. She would later tell the inquiry she did not believe the European standards were relevant in the UK market. Asked if she did not tell customers about the new fire rating in order to avoid 'damaging her sales', she answered 'absolutely not'.[11]

'I THINK I OWE YOU BOTH EITHER LUNCH OR DINNER AT SOME POINT'

We will come to the construction job that fitted the cladding to the walls of Grenfell Tower shortly, but for now it is necessary to introduce the lead architect in its early years: Bruce Sounes, from Studio E. This would be Mr Sounes' first residential cladding project. He turned to the internet to do early research and found CEP Architectural Facades, which cut and sold cladding panels, and arranged a meeting with its director Geof Blades in March 2012.[12]

At this point, Mr Sounes wanted to use zinc, due to its aesthetic appearance. He had not made any consideration of fire performance, but he had by chance specified an entirely non-combustible metal panel. He thought it looked nicer and would be less prone to rust. But what he did not know is that Mr Blades' company had a commercial relationship with Arconic and primarily cut and sold its ACM. In October, Mr Blades introduced Mr Sounes to Debbie French. She began to set out the options for ACM, telling him it would be a cheaper solution for his project.

And this was a factor which would weigh heavily. As we will see, the job was above budget and changes were required in order to cut costs. The zinc cladding was switched for ACM. Debbie French sent over the BBA certificate that was taken as evidence that the product complied with requirements. This change was then approved by the RBKC planning committee, who insisted on cassette panels despite the additional expense they imposed out of a fear that rivets would become 'rust attractors' and then stain the cladding. A petty concern with aesthetics resulted in the most dangerous form of this already dangerous product being selected for the tower.

The team refurbishing the tower ordered some ACM samples to install a mock-up. A representative of cladding subcontractor Harley emailed CEP's Mr Blades and Ms French to let them know both would be in line for big sales. 'All I can say is that you will be taken out for a very nice meal very soon,' replied Mr Blades. 'Thank you for your hard work and perseverance in putting Reynobond forward,' added Ms French. 'I think I owe you both either lunch or dinner at some point.'*

'WE ARE IN "THE KNOW"'

At around the same time, the BBA was trying to carry out a periodic review of the certificate it had first issued to Arconic in 2006. It wrote to Arconic seeking written confirmation that there had been no changes that would invalidate the certificate. This was important. The certificate was out of date and wrongly said the panels were Euroclass B when they were now 'E'. But Arconic ignored multiple emails all the way through to January 2015 when the BBA finally gave up. The BBA could have suspended the certificate at this point, and issued a public notice saying it was no longer valid. It did not.

Instead, noting 'problems getting in touch with you', the BBA told Arconic it had 'decided to go ahead with the information already in hand'. This meant, in practice, material found during a search of Arconic's website. By now, Arconic's website only listed the 'FR' product as

---

* Inquiry transcript, 10 September 2020. All three have denied any incentives were given to put forward Reynobond products and described their relationships as merely professional. They have said no meal was ever had.

Class B. But this was enough for the BBA to reissue its certificate with the now misleading claim.*

The building inspector who signed off Grenfell as compliant would rely on this certificate as evidence that the cladding complied with regulations in early 2015. The panels – despite testing which showed them to be violently combustible in a fire almost eleven years earlier – were sold, installed on the walls of 129 homes and signed off as compliant.

Meanwhile, in France Mr Wehrle was becoming increasingly concerned as ACM fires continued to strike. Following a fire in Strasbourg which neighboured a tower block clad with ACM he emailed five senior staff members. 'We were very lucky,' he wrote. 'Fortunately the wind didn't change direction but we really need to stop proposing PE in architecture. We are in "the know" and it is up to us to be proactive... AT LAST.'[13]

By May 2016, Arconic did finally change tack – at least in France. In an email to its French salespeople, a senior member of the team wrote that due to 'the very significant difference in heating capacities' between the PE and FR cladding they should 'systematically confirm in writing' the 'requirement' for FR on all projects going forward.

---

* Inquiry transcript, 15 March 2021. The BBA project manager Valentina Amoroso said she assumed the CSTB (the main French research and testing centre for construction) would have disclosed any new tests and believed Arconic were contractually obliged to provide them. Arconic believes the certificate only applied to the panel in its unfabricated form, and therefore argue the certificate was not misleading.

But such a warning was never issued to the UK market. Asked why, Mr Schmidt said he 'could not explain it', other than a possibility that the company understood the French regulations more clearly than those in the UK.* The UK remained a 'PE market' in the eyes of Arconic and would remain so until the cladding at Grenfell burned in the way testing had suggested it might thirteen years earlier.

Why was Arconic able to sell a dangerous product into the UK market for so long? For its part, Arconic claims that its product was not inherently risky. Instead, it said the risk depended on its use and it was perfectly possible for it to be used safely. It says that responsibility therefore lies with those who elected to use it on Grenfell Tower in the manner that they did and those who failed to maintain the condition of the tower in the months and years prior to the fire. It also defended its BBA certificate – insisting that since it sold the panels as flat sheets for others to bend into shape, the fact that it had a lower fire performance when shaped into cassettes did not mean the certificate was invalid if applied to the basic product. It said the evidence of other buildings with ACM reveal it was widely used in the UK, because it was accepted within the regulatory regime – and Arconic was entitled to rely on this as safe.

A lack of transparency permitted Arconic to profiteer from an obvious fire hazard. Nothing existed to force

---

* Inquiry transcript, 15 February 2021. Arconic said in its written closing that training given to UK staff did reference the 'E' classification for ACM PE in cassette form.

it to release the crucial testing that exposed the danger, particularly of the cassette panel. Were the French testing facility that carried out the 2004 test, for example, obliged to release these findings, the dangers of the product would have been known much earlier and made available to the BBA. Instead testing of this kind is branded 'commercial confidential' and the intellectual property of the company which carried it out. If they decide not to release it, it will never be released.

But the UK government cannot simply react with shock. It could have set higher standards. Political and economic choices saw us adopt lower standards than Germany and others around the EU. We wanted to remove barriers to doing business, and in doing so became a market where dangerously combustible cladding could be sold.

Equally, Arconic cannot blame the UK government's indifference to fire safety. It had its testing from 2004 which demonstrated the risks, and all the warnings from senior members of the company. But corporate interest took precedence over human morality. Corporations will act in the interests of profit. When law no longer holds them to account, we have no choice but to wait for the consequences.

7

1.45 A.M.

As the architectural crown ignited, the fire at Grenfell began to spread out. Flames moved clockwise around the corner of the building, and anti-clockwise across the face where it had begun. As this happened, molten plastic flowed down the tower, starting new fires which then spread back up. Two long columns of flame spread out around the building, moving steadily in opposite directions towards the windows where the silhouettes of those trapped were visible from the ground.

As flames spread around the top of the building, conditions for those sheltering in the flats on the twenty-third floor quickly got worse. At 1.45 a.m., Sakina's son phoned her and she told him they were trapped, there was smoke in the flat and she was struggling to breathe.

In the flat where Debbie Lamprell had fled the conditions were even worse. At 1.41 a.m., Debbie began a forty-minute call with a fire service operator. The operator recalled that Debbie was 'screaming down the phone' as soon as she picked up.

'We're on the top floor at Grenfell Tower. Please help us,' Debbie said. As she tried to give her postcode and location, Debbie was coughing and often inaudible – complaining of the 'thick smoke' which was already filling the flat where they were sheltering. She said the fire was 'coming up' the walls, but the smoke was already pouring in through the windows.

The call handler advised her to put something under the doors to prevent the smoke coming in – but Debbie said that it was coming through the windows and could not be stopped. 'Can't breathe,' she said. 'I'm going to die up here.'

'You're not, you're not. Listen. Is there a lot of smoke?'

'Yes. Thick, black.'

The call handler assured her the information would be passed on to the crews and they would be rescued. She said she lived in Flat 161, but had fled to Flat 201. Nonetheless, her location was initially recorded as Flat 161 at 1.46 a.m., before being amended to Flat 201 on the twenty-third floor six minutes later. But this change was not communicated to the officers on the ground at the tower. Not long afterwards a crew was dispatched to find her and the group she was with in Flat 161. They would battle up the tower, reach the nineteenth floor, and find the flat empty.

1.50 A.M.

By 1.50 a.m., twenty-two fire engines had arrived at the scene carrying 114 firefighters. Michael Dowden – with the relatively junior rank of watch manager – was still in charge, despite at least two officers of a higher rank being present. This would be branded 'a strikingly unsatisfactory feature of the incident' in the report of inquiry chair Sir Martin Moore-Bick.[1]

He was feeling out of his depth at this point. 'I don't know what the fuck happened,' he said to another officer who arrived and went to speak to him.

The situation was grim and getting worse – beyond the scale of anything seen in London since the Blitz.

The control room had received twenty calls from residents trapped inside the tower, ranging from the third floor to the very top of the building and on different sides. But Dowden had not spoken to the officers in the control room and had not been passed any information about the volume of calls coming in. Another (more senior) officer had been put in charge of co-ordinating this information. By this point, in Sir Martin's view, full evacuation was 'the only realistic way of minimising loss of life'.[2] Instead, Dowden's focus remained on fighting the fire which – by this stage – was an unwinnable battle. Sir Martin's view was that Dowden's failure to order an evacuation was because 'his training did not equip him' to make this decision, not because of incompetence on his part.[3] He was left feeling helpless as he watched the fire spread.

Shortly before 2 a.m., he finally handed over command to station manager Andrew Walton. Their conversation was brief, and Dowden's information about conditions inside the tower was limited. There was too much traffic on the radio to get reliable messages through. Walton sent him into the tower to collect more information about the progress of the fire inside the building. A few minutes later, an even more senior figure – deputy assistant commissioner Andrew O'Loughlin – took over command from Walton.

O'Loughlin had been shocked by what he'd seen when he arrived at the tower. 'The grassy area underneath the east elevation was getting covered in huge chunks of debris, probably up to a foot in depth. Some of this debris was six to ten feet in size, the size of a patio door, and a lot of it was alight,' he said in his witness statement.

While he had heard of several calls from residents trapped within the tower via his radio, he had not fully appreciated the extent to which the fire was breaking back into the tower and putting residents' lives at risk.[*] He took no steps to lift the stay put advice and evacuate the remaining residents. He made for the command unit.

*

On the nineteenth floor, Pily and Nick were sheltering in Flat 165. They were the only ones left on that floor. Nick had already seen the lobby outside their home, full of smoke. He was aware it was the biggest threat to their safety and had shut the door very quickly to prevent it getting into the flat. He placed a wet towel at the base of the door to help keep the smoke out, and left three more soaking in the bath for himself, Pily and their dog – should more of the smoke get into their home.

---

[*] His evidence varied on this point – having recorded in notes in July 2017 that there was 'fire in flats' when he arrived, he later told the inquiry he thought the cladding was only burning externally and had only got into a few flats through open windows.

Pily was resting on the sofa. Nick knew she couldn't make it out of the building without help and their only option was to wait for rescue from the fire service. At that point Nick's phone began to ring. On the line was his fellow dogwalker from the estate. He was outside the tower and could see the progress of the flames which had spread up the building and were starting to spread around it. He was panicking, urging Nick – who was still quite calm – to flee the building immediately. He still believed though, in part considering Pily's condition, that the safest thing to do was wait.

At 1.56 a.m., Nick made his first call to the emergency services. 'Oh, hi. Good evening,' he said when the operator picked up. 'I think you know about it already, we're in Grenfell Tower. The whole tower's on, on fire, and we're on the nineteenth floor.'

He explained that he was with his wife and dog and that smoke was filling the landing outside their flat. The operator told him the firefighters would be told where he was and 'would be coming door to door to make sure everyone's safe'. Despite the fire by now having reached the top of the building, the operator assured him that it was 'actually on the fourth floor'. 'If anything changes give us a call back, okay?' the operator said. 'But, yeah, stay where you are. All right then? Thank you.' By now, from his window, Nick could see smoke and ash floating past in the night sky and the glow of the fire from the other side of the building.

On the eleventh floor, Natasha Elcock had got dressed and woken her daughter up. She'd phoned 999 twice already and asked for someone to be sent to rescue them. Her partner had stepped outside once already to see if it was safe to leave but had quickly returned to the flat and said the landing was not passable. He thought he had seen a blue flame on the landing. She went to the bathroom, put the plug in the hole and turned on the taps. Her partner asked her what she was doing. 'Well the house is going to be fucked, a bit of water ain't going to hurt,' she replied. She called the fire service again and told them smoke was starting to seep into their flat.

*

As 2 a.m. approached, Debbie was still on the phone to the emergency services. She would carry on speaking for another thirty minutes. By then, the fire had got in through the windows of the flat and those sheltering had barricaded themselves inside a bedroom to await rescue.

'I've told them where you are, they know where you are. They will come and get you. All right? They need to try and find a safe route to you but they will get to you, okay?' the call handler told Debbie.

'I don't want to die.'

'You won't. You're not going to die.'

'Please, the smoke is so thick.'

Some of the other people in the flat began vomiting from the smoke. Then there was a loud cracking noise. The fire had broken the bedroom window. Debbie told the call handler the room was on fire and the call records the sound of her trying to break down the door to escape. 'It's too hot,' she said. Her replies gradually became inaudible. And then silence.

8

'THE BENEFITS IN TERMS OF LIVES SAVED…ARE CONSIDERED TO BE LOW'

Behind the combustible cladding panels on the walls of Grenfell Tower was tonne upon tonne of combustible foam insulation. This was far from uncommon.

Selling insulation is big business. The global industry is expected to grow to $70bn by 2024, as regulations tighten, thanks to efforts to combat climate change and use less heating.

The insulation industry has lobbied hard for the minimum standards and funding streams which underpin this work. As *Sky News* reported a few months after Grenfell, the UK lobbying arm of the plastic insulation industry boasted of its ability to 'influence UK and local government, specifying authorities, relevant approval and certification bodies' and having 'high level involvement in the drafting and regular revision of British and European standards [and] the Building Regulations'. Rob Warren, a former technical director at Celotex, told a trade publication in 2015 that regulatory change was 'the greatest driver of plastic insulation sales'. 'You cannot give insulation away and the public are not really interested,' he said.[1]

There are two main types of insulation available on the UK market: rigid, plastic foam boards and rolls of non-combustible wool, woven from rock or glass fibres. These two sides of the industry are huge, powerful and engaged in an unending and bitter war to dominate. The plastic boards are thinner, lighter and offer higher insulation performance. That gives them an enormous advantage over their mineral wool rivals. But their Achilles' heel is fire.

Plastic is always combustible. And not only will it burn, it will also release large quantities of toxic smoke when it does. Plastic insulation manufacturers have engaged for a long time in marketing campaigns to downplay these risks. Promoters of these products insist that they 'char rather than burn'. But as one fire science academic explained to me, this is misleading: 'Charring is burning… It's marketing speak to say anything else.'[2] I have seen marketing literature describe the fire performance of combustible phenolic foam as 'exceptional', crediting it with 'zero or very low flame spread with negligible smoke emission' and insisting it is 'capable of meeting or exceeding all international building regulation requirements'.[3]

Marketing is just one arm of this effort. The firms have also participated directly in the process of making fire and smoke regulations. Obscure standard setting bodies which produce the guidance are dominated by industry lobbyists. 'Over the years I've been there it's gone from having experts there to having nearly all the people present lobbying for the plastics industry. They are all lined up to try and block things that threaten their product,' says one source who has attended some of these committees. 'This happens in every industry,' he adds. 'But when it comes to life safety, it really shouldn't be the case that the people writing the standards are the ones selling the products.'

In 2004/5, the BRE wrote a report for the government considering the necessity of introducing smoke toxicity limits for materials used in internal walls and ceilings. All European countries bar the UK and Ireland, the report said, had some sort of standard. But the UK would not be following suit. The report also said the standards would 'have a significant impact on product sales'. 'The most demanding option could potentially affect sales with an annual value upwards of £249m.' This was weighed against the impact on safety. 'The benefits in terms of lives saved or reduced injuries… are considered to be low,' it reads. 'Using accepted valuation techniques… the annual benefit is estimated to be £174,000 per year.'[4]

On Grenfell Tower, the majority of the insulation was polyisocyanurate which had been sold by Celotex, while a smaller amount was phenolic, sold by Kingspan. Both were combustible, and both let off toxic smoke when burned. Just one kilogram of the Celotex product was enough to fill a normal sized room with an 'incapacitating and ulti-mately lethal' level of smoke when burned according to a 2019 study. There was an estimated 19,650kg on the walls of Grenfell Tower.[5]

And Grenfell was not unusual. Similar insulation products were in use on thousands, perhaps tens of thou-sands, of high- and medium-rise buildings around the UK. Materials prone to release poisonous smoke when burned are commonplace on the walls of our homes.* They started their ascent in the 1990s.

'ONE REGRETS, THERE ARE NOW COMMERCIAL PRESSURES'

In that decade a series of fires at commercial warehouses built out of 'sandwich panels' – which held the insulation between metal sheets – led to huge blazes which destroyed whole properties. The fires were difficult to extinguish because the burning insulation was protected by the metal, and the entire building was often lost. In total, throughout the decade, twenty major fires occurred in commercial buildings with combustible insulation, including the Sun Valley factory in Hereford where two firefighters were killed in 1993.[6]

'Often the heat generated was phenomenal,' says one source who worked in a technical role for the Fire Brigades Union (FBU). 'It got to the point where you were saying you can't commit firefighters into one of these buildings.'[7] In 1997 an edition of *Fire*, a journal for the fire protection profession, was guest edited by the FBU. It called on the new Labour government to impose tougher restrictions

---

* All materials that burn release carbon monoxide, but only those containing nitrogen also release cyanide. Cyanide was present in the smoke at Grenfell Tower as a result of the insulation panels, window surrounds and upholstered furniture burning inside flats. A toxicology expert said the gas likely had a 'minimal' impact on the cause of death, which was primarily carbon monoxide inhalation, although he said it was possible the cyanide caused some victims to collapse faster.

on the use of these products. 'We sincerely hope, now we have a new [Labour] government in place, that they will act decisively… before another fire tragedy possibly involving serious loss of life occurs,' they wrote.[8]

As we discussed in Chapter 2, by the end of the 1990s, rising concern about cladding fires resulted in an investigation of the risks by a Parliamentary select committee. The FBU advised the committee to ban the use of all combustible materials on tall buildings.[9]

But MPs also heard from witnesses from the BRE. As we've seen, the organisation – historically the British state's research centre – had been privatised in 1997. It now required commercial income to fund its work. And it had been developing a large-scale cladding test since the 1991 fire at Knowsley Heights. Internal documents note 'a great deal of interest' from manufacturers and say 'there appears to be a large potential for commercial funding' for the tests.[10]

At this point, regulations for rainscreen cladding systems allowed Class 0 cladding, but required insulation to meet the higher standard of limited combustibility. What the BRE emphasised was the need to move away from unreliable small-scale testing and test the full cladding system. While the logic of this seems compelling, it was also misleading. It was the Class 0 test that was unreliable. Testing to make sure a product was not combustible worked just fine.

But the BRE's advice suggested the problem was small-scale testing per se, not one outdated and dodgy test. They urged the committee to recommend large-scale tests as a replacement. Their witness, Peter Field, was candid: the BRE expected to make money from this work. 'Clearly, in days gone by, when we were part of [the DoE] then this work… would have been done in the public interest,' he said. 'One regrets, there are now commercial pressures that require clients to place formal contracts with us before we can undertake work.'[11]

The committee ultimately combined the recommendations of the FBU and the BRE. It said all cladding systems should either be entirely non-combustible or pass one of the new tests. But the government only half accepted this recommendation. It left the flawed standards in place, which required cladding panels to be Class 0 and insulation to be limited combustibility. But it also introduced the testing system as an alternative.

This represented a dangerous back door. Combustible plastic insulation products were locked out of the high-rise market by the limited combustibility requirement. But now, if the plastic insulation manufacturers could find a system which passed a large-scale test, they could start selling their product for use on high-rises.

And at this point, the money involved in insulation was about to snowball. The UK was a signatory to the Kyoto Protocol on climate change, which would soon be coming into force, resulting in tougher requirements to insulate large buildings. Both Kingspan and Celotex smelled profits. 'With the government committed to the Kyoto Protocol… the building regulations will be amended in the spring of 2006,' Celotex wrote in its financial statements for that year. 'Celotex Limited expects significant growth as a result.'[12]

'Kingspan's range of products are well positioned to benefit from the general thrust of this directive,' Eugene

Murtagh, chair of Kingspan, wrote in the company's annual report in the same year. 'The pressure to transform is set to increase in the coming years. Kingspan intends to influence this transformation, and is well placed to take advantage of all these developments.'[13]

But to win the forthcoming jobs for high-rise buildings, they needed to pass a test.

'A RAGING INFERNO'

The large-scale test the BRE had developed is known in the trade as a 'BS-8414 test' after the guidance note which sets its parameters. It involves building a 9m-high mock-up of a cladding system, lighting a fire in a crib at its base and monitoring for an hour. If temperatures above 600°C are recorded inside the first 15 minutes, or if flames spread above the top of the rig at any point, it fails. Avoid these two outcomes and it passes.

This test has its limits. With no windows, openings or wind, the fire behaves more predictably than it does in the real world. Only monitoring the temperature for fifteen minutes also means systems that burn fiercely but slowly can pass. And the failure criteria is limited: large pieces can fall off and the fire can produce large volumes of toxic smoke without the system failing.

Kingspan carried out its first test in May 2005. The system it tested used its flagship K15 insulation behind a wall of heavy cement fibre cladding panels. This was a strange choice of material. Cement fibre is not used in this way in the real world. The panels are too heavy to lift and fix to tall buildings and they are not generally weatherproof. So what was the point of the test? Cement fibre is non-combustible and highly fire resistant – meaning it protected the insulation better than some more common products and gave Kingspan a better chance at a pass.*

Emails show that Kingspan was hoping to receive an 'extended field of application' report from the BRE, effectively clearing the use of its insulation for all cladding systems with a non-combustible exterior panel on the basis of this single test. But the BRE declined to provide this assurance. It told Kingspan the test was for the exact system only.[14]

This is not how Kingspan marketed it. A flyer produced shortly after the test appeared to offer a sharp warning about the dangers of combustible insulation. 'The last thing you need behind a rainscreen cladding system is flammable insulation,' it said. 'A careless cigarette, an act of arson or an electrical fault could all be sufficient to start a fire. Combine that with the chimney effect of a ventilated cavity and you could be looking at a towering inferno scenario in no time. Unless, that is, you have had the foresight to install an insulation material that will limit the spread of fire.'[15]

K15, the flyer implied, was one such product. It said the insulation had been 'successfully tested to BS8414' and was 'acceptable for use above 18m'. It did not mention that this was true for only one specific system – where it was used in combination with cement fibre cladding panels.

---

* Kingspan says the test was a 'bona fide test of a generic cladding system' and 'entirely appropriate for its time'.

The flyer also emphasised that the product 'achieves a Class 0' rating. While Class 0 was not relevant to the use of insulation, only applying to the external cladding panels, not all builders knew this. And the fact that it had this rating sounded reassuring to anyone worried about fire. But did it really possess this rating? Kingspan had in fact obtained it by ripping off the foil which was normally attached to the product and testing it in isolation. During the inquiry, the barrister questioning Kingspan witnesses compared this to stapling a foil facer to a stick of dynamite and claiming it obtained Class 0 – a comparison the witnesses rejected. In 2016, two Kingspan employees discussed this testing via text: 'Doesn't actually get Class 0 when we test the whole product tho. LOL!' 'WHAT. We lied?' his colleague replied. 'All we do is lie in here,' the first replied.*

With these marketing claims, the firm began targeting and winning jobs on high-rises.

Soon after, Kingspan made some technical changes to the way it made K15. The foil facing on the front of the sheet was perforated, and the chemical formula for the plastic was altered so that it retained more of the 'blowing agent'. The hope was to improve its thermal performance in an increasingly competitive market.

In December 2007, Kingspan sought to pass a different version of the large-scale test it had completed successfully

---

* Inquiry transcript, 9 December 2020. Asked about this exchange, a senior Kingspan manager called it 'very disappointing' and said it was 'not true' to say all the company did was lie.

in 2005. This test used non-combustible solid aluminium cladding panels in front of the new technology K15. It was a disaster. An internal report prepared by Kingspan staff member Ivor Meredith described the rig as 'a raging inferno'. 'The phenolic [insulation] was burning on its own steam and the BRE had to extinguish the test early because it was endangering setting fire to the laboratory,' he wrote.[*]

Under a heading that said 'Why did it fail' he wrote that the new technology insulation was 'very different'. Whereas the old technology had 'turned into a light ash and fallen away' the new product 'burned very ferociously'. He noted 'unofficial' comments from the BRE that the insulation was 'fully involved' in the fire and 'continued to burn when the flame source had been extinguished'.[**] Kingspan insists, despite these comments, there is 'no scientific basis' for saying the new technology had a lower fire performance. It claims the failure was a result of the system tested, not the insulation – pointing out that a similar system still failed, even with non-combustible insulation.

Nonetheless, it did not release this testing, even to other arms of its own business. A subsidiary company 'Kingspan Offsite', which sold whole cladding systems, ran its own

---

[*] Inquiry transcript, 23 November 2020. Kingspan's lawyer played this down, citing a witness who did not agree the fire was 'spectacular', and arguing there was no clear evidence that the changes reduced the fire performance. They say this test also fails in a similar fashion with non-combustible insulation.

[**] Several BRE witnesses have denied making 'unofficial' comments of any kind to Kingspan, saying it was not the BRE's practice.

tests on a system containing the new K15 in summer 2008. They also experienced rapid and serious failures. They asked the technical team if it was 'normal for K15 to continue to burn for in excess of 30 minutes from the removal of the ignition source?'[16]

But they were given no indication that the product had changed, or that a prior test on it had produced a similar result. Instead, the technical team said there were 'always inconsistencies' in fire tests and added that the cause of the failure was 'difficult to determine'. 'I'm spinning so hard I'm dizzy!' wrote manager Philip Heath in one internal email.*

'I THINK [THEY] ARE GETTING ME CONFUSED WITH SOMEONE WHO GIVES A DAMN'

Like Arconic, Kingspan approached the BBA to verify its claim that the product could be used on tall buildings. The BBA was told nothing of the changes to K15 and the subsequent test failures. Kingspan simply provided the report on the 2005 test, as well as a good deal of Kingspan marketing literature about the good fire performance of its product.

Kingspan got the certificate it wanted. First released in October 2008, the certificate said the product had Class 0 fire certification and would not contribute to 'the development stages of a fire or present a smoke or toxic hazard'.

---

* Inquiry transcript, 30 November 2020. Mr Heath told the inquiry this was a reference to being busy and trying to 'create a delay' to allow time for a new, safer product to come through.

But this was a problem. It emerged at the inquiry that the BBA had based its certificate on testing on different products and its former deputy chief executive accepted the reassurance was 'inaccurate and misleading'. The certificate also said 'the product' had met the necessary criteria during a large-scale test, and encouraged builders to seek advice from Kingspan if they wanted to use it over 18m. This too was a problem. The large-scale test was never designed for a single product, but for an entire system. The rules only allowed its use if builders reconstructed the 2005 test system exactly. But this certificate was merely advising them to phone Kingspan and check – opening the door for its marketing team to green light the use of the product in various other systems.*

Internally, Kingspan batted away queries from those who were concerned it might not be suitable. 'I think [they] are getting me confused with someone who gives a damn. I'm trying to think of a way out of this one, imagine a fire running up the tower!!!!!!!!!!!!', wrote Mr Heath internally in response to one query. Discussing a consultancy which was querying the use of the product on high-rises, he said they 'can go fuck themselves and if they're not careful we will sue the arse of [sic] them.'**

---

* Kingspan claimed the certificate was not misleading – referring to comments from another BBA witness who said it would have been understood properly if read by 'a suitably competent reader'.

** Inquiry transcript, 30 November 2020. Confronted with these emails at the inquiry in 2020, Mr Heath apologised for the 'unprofessional' language and said he was feeling stressed due to a friend's terminal illness.

The firm also secured certification from an organisation called 'Local Authority Building Control' (LABC), a group which represents inspectors at local councils who sign off projects as compliant. As such, securing a statement from them that the insulation was suitable for high-rises would carry a great deal of weight with the inspectors. Their certificate, published in May 2009, said the product could be 'considered a material of limited combustibility' and was 'suitable for use' in parts of a building 'more than 18m above the ground'.*

This was completely wrong: K15 was not a material of limited combustibility. But when Kingspan saw a draft of the document, Philip Heath responded internally with a single word 'FANBLOODYTASTIC'. When a senior manager asked him how the certificate had been secured, he wrote: 'We threw every bit of fire test data we could at him, we probably blocked his server. In the end I think the LABC convinced themselves [K15] is the best thing since sliced bread. We didn't even have to get any real ale down him!'**

With these two certificates in the bank – Kingspan's marketing team were able to secure job after job on high-rise buildings. Seeking the cheapest way to attain required

---

* Inquiry transcript, 21 March 2021. Kingspan stresses that it was not responsible for the wording of the certificate.

** Inquiry transcript, 1 December 2020. Mr Heath says he was aware K15 was not a product of limited combustibility, but did not think the certificate was misleading. He said he believed at the time that it only applied to the material's use in a particular system but could see 'with hindsight' why it may have been misleading.

insulation standards, builders combined the combustible K15 insulation with all manner of combustible cladding panels. Unnoticed by most, we were stuffing our walls with combustible plastic foam.

'IN THE EVENT OF A FIRE IT WOULD BURN'

In 2010, the rewards on offer to insulation manufacturers were raised. Following sustained lobbying the building regulations guidance was strengthened again, which meant even more insulation would be required on new build homes and retrofit projects. The higher standards favoured plastic, which was lighter and cheaper and could achieve greater 'u-values' with thinner boards.

As part of a campaign called 'Insulating Britain', Celotex booked a bus tour around the country where a large pink double decker plastered with its branding announced 'The building regulations are changing. Don't just insulate it, Celotex it!'[17]

But the firm was locked out of one corner of the market. Unlike Kingspan, it had not passed a test and had no way to persuade builders it was suitable for high-rises. This was a small but significant part of the market. Celotex estimated it to be worth around £10m a year to their business. In 2012, Celotex – a medium-sized UK company – was sold to Saint Gobain: a multinational French materials firm with global turnover approaching £40bn. To boost Celotex's profitability, Saint Gobain wanted it to develop and sell new product lines. It set Celotex a target:

15% of its profit increase had to come from new products. Celotex's senior staff were hauled to meetings in Paris to report on progress. An obvious target was high-rises.[18]

Jonathan Roper, a business graduate from the University of East Anglia, aged just twenty-two and in his first job, was put in charge of researching how Celotex could break into this market. Mr Roper's market research quite quickly revealed how Kingspan was winning these jobs. 'An architect will be told that K15 is applicable for above 18m…and that suffices from their perspective,' he wrote in an email in November 2013. 'Contractors opt for more cost effective solutions and although they are liable for what goes into that building, they do not know enough about the fire test to challenge.'

However, Mr Roper was not sure Celotex should follow this route. 'Do we take the view that our product realistically shouldn't be used behind most cladding panels because in the event of a fire it would burn?' he wrote.

His managers disagreed. The commercial team at Celotex were keen to replicate Kingspan's approach. Mr Roper arranged a test at the BRE's burn hall for February 2014 using a cement fibre cladding panel. This test failed. The cement panels cracked from the heat of the fire, the flames entered the gap between the cladding and insulation and ripped through the system. But Celotex was still keen to keep trying.

They ran another test in May with a thicker panel they believed would be less likely to crack. This time, they reinforced it with fire-resistant boards around the temperature monitors. The rig was built in full view of BRE staff and the panels were not well hidden: they were slipped behind boards of a different thickness and colour and appeared in photographs taken by BRE staff before the test and afterwards when the rig was dismantled.*

Mr Roper claimed there was then a 'heated debate' at a key meeting, with Celotex management ultimately deciding that revealing the presence of the panels would 'limit sales'.** This upset Mr Roper, but we know he didn't know what to do. He would later accept at the inquiry that the company's actions were 'deliberately misleading and dishonest' and 'a fraud on the market'. He said he had been made to 'lie for commercial gain'. 'I still lived with my parents at the time and mentioned that to them that I felt incredibly uncomfortable with what I was being asked to do,' he later told the inquiry. 'I went along with a lot of actions at Celotex that, looking back on reflection, were completely unethical,' he added.***

Celotex then produced a document which said the product had passed the test and was 'suitable for buildings

---

* Mr Roper and another Celotex witness have claimed the BRE were aware this was being done. The BRE's burn hall manager has firmly denied he knew about the addition of the boards.

** Mr Roper was not at the meeting and his evidence of this was second-hand.

*** Inquiry transcript, 16 November 2020. Celotex claims these issues were unknown to its current management and when they emerged after the fire, they were reported to trading standards, police and publicised by the firm on its website. The employees involved who remained at the company were subject to disciplinary proceedings and have all now left.

above 18m' on the banner of every page.* The details of the test were relegated to small print and no mention of the additional fire barriers was included at all. It also obtained a certificate from the LABC which said the insulation could be used 'with a variety of cladding systems' and that its test pass meant it is 'therefore acceptable for use in buildings above 18m in height'. Now it could compete with Kingspan for high-rise contracts. And there was a big one coming up in West London.

In August 2014, a Celotex salesperson emailed Harley Facades, the sub-contractor working on the Grenfell Tower project to say RS5000 had passed the test and was therefore 'acceptable for use on buildings above 18m'. In February 2015, it offered the firm a 47.5% discount to use the insulation on the tower. It would later ask for permission to use Grenfell Tower as a 'case study' for the product – one of the first high-rises it had ever been installed on.[19]

'AN ACCIDENT WAITING TO HAPPEN'

While all of this was going on, concerns about the use of combustible insulation were swirling in the industry. Facade consultants challenged the idea that products such as Kingspan and Celotex were compliant for high-rises.

---

* Celotex claims competent professionals would have understood this to be within the context of the test it had passed, which was mentioned in the booklet. An expert witness to the inquiry agreed that competent professionals should have read the literature carefully and understood it.

These concerns reached the ears of government. On 2 July 2014, Brian Martin emailed a senior figure at the National House Building Council (NHBC) saying that he had been 'talking to a few folk about fire safety and facades recently'. 'Allegedly – several buildings have been erected where [plastic] insulation has been used in cladding panels well over 18m in height… Again, allegedly, many of these buildings are blocks of flats.'

'The purpose of my email is a friendly warning,' he added. 'You might want to double check with your inspectors and plan checkers that they are on top of this.'[20]

The NHBC is the country's largest private 'building control' inspector. Its inspectors are employed by builders to sign off their jobs as compliant with the regulations and it also provides warranties on them once built. What Mr Martin did not know is that it had, by this point, already been warned about the use of Kingspan insulation on high-rises and had asked the firm to provide fresh testing data to back up its claims. This was of major concern to Kingspan. If the NHBC started rejecting its product, it would lose millions. But the stakes were also high for the NHBC - it had been approving Kingspan for years, meaning it had a long liability for its use. And its clients (builders) liked using the cheap, efficient product and would not react well to being told it was suddenly banned. One staff member cautioned internally that doing so would 'cause a major issue with our customers [housebuilders]', as they would 'be forced to use Rockwool'. This was despite senior NHBC staff having heard through the grapevine about the failed tests on systems containing K15.

The NHBC's head of technical operations, Steve Evans, wrote back to Mr Martin outlining the issue with Kingspan, but assured him 'there is no reason to suspect buildings built with Kingspan K15 are at risk at this time. It is just the fact that the testing carried out to date does not bear this out'.[21]

In an email to government colleagues, Mr Martin explained the consequences of what was being discussed. 'It would appear that a number of tall apartment buildings have been insulated using combustible insulation materials when perhaps they shouldn't have been,' he wrote to colleagues in the department. 'If it is a problem, some blocks may need to have their cladding replaced. (possibly a lot of them).' So it would prove – but it would take the Grenfell Tower disaster for these checks to be made.[22]

Meanwhile, Kingspan had been trying to secure fresh testing to satisfy the NHBC's concerns. Its first two efforts had failed but in July 2014, it secured another pass. 'F*cking happy days!' wrote one Kingspan employee internally. 'I think we have just pissed over [rival manufacturers] Knauf, Xtratherm and Rockwool!'[23] But something was amiss. The insulation tested was not the same as the product on the market. Instead, it was a new research and development product that had a foil facer double the thickness of the one on the market and a different chemical composition. Kingspan, once more, was using test passes on systems containing products different from the one it was selling to justify its claims about safety. Nonetheless, the new test pass was not enough for the NHBC, who wanted further assurance that it could legitimately be used in a range of systems. Kingspan sought the opinion of respected industry expert, Dr Barbara Lane of Arup, who would go on to be an expert witness at the Grenfell Inquiry. They hoped she would endorse its product. But she would not oblige them. Saying she was 'deeply concerned' about the 'lack of understanding' in the industry, she warned NHBC that 'the use of highly combustible materials in residential buildings is now simply an accident waiting to happen.'[24]

The NHBC had now heard about the potential dangers of K15. It had not got the additional testing it had asked for. It could have simply rejected the product and insisted on non-combustible insulation instead. It did not. Instead, as part of an industry group of other organisations involved in building control, the NHBC helped draft a guidance note which said that 'if no actual test data exists' for a product, builders could submit a 'desktop study report' from an expert. This guidance was first issued in June 2014. From now on, insulation manufacturers and builders no longer needed a test for their product to be used on high-rises, they merely needed a piece of paper.[25]

It appears that, despite this, the NHBC did come close to rejecting K15. In early 2015, NHBC wrote to Kingspan, copying in its chief executive Gene Murtagh. It said it was going to start rejecting K15 for high-rises and would 'inform our builder customers of our concerns at the earliest opportunity'.

Kingspan's response came directly from its solicitors. It warned that the NHBC's action would cause Kingspan 'very significant financial loss' and 'constitute negligent misstatement and defamation'. NHBC moderated its position. Rather than rejecting the product outright, it said that it would accept the insulation so long as a desktop study was provided.[26]

And the position would get looser still. In July 2016, the NHBC published guidance setting out various systems which it considered were acceptable, without even needing a desktop study. This included both Celotex's RS5000 and Kingspan's K15 insulation. It also included ACM panels, provided they had a Class B rating. The rules had been almost completely undermined. While NHBC did not sign off Grenfell Tower, research by *Inside Housing* revealed it had signed off more than fifty projects with dangerous ACM systems.[27] With our system of privatised testing, certification and sign-off wholly failing to enforce the rules, combustible insulation was being installed on thousands of tall buildings around the country.

One of these buildings was Grenfell Tower. When the supply of Celotex RS5000 slipped in spring 2015, Harley Facades substituted it with a quick order of K15. The internal team said they reviewed certificates and assumed the product was compliant.[28] Alongside Celotex, it was placed in a system which had never been tested and was not even subject to a desktop study. The accident Dr Lane had predicted was looming.

'CROOKS AND KILLERS'

Kingspan has argued firmly at the inquiry that its insulation was in no way to blame for the fire at Grenfell Tower. It emphasises that only a small amount was used – and that the inquiry has concluded the primary reason for the fire spread was the combustible ACM panels. An expert witness has said the insulation in the cladding system only contributed to between 2% and 10% of the burning and using non-combustible insulation would have made little difference to the speed at which the fire spread.[29]

The firm claims its product is safe to use when installed correctly in systems with compliant materials, and that the best way of finding systems which are safe to use is to test full systems. It has legitimately passed several tests with K15 since the fire, and insists these demonstrate that its product is safe if installed correctly. It has apologised for what it calls 'the unacceptable actions of a small group of former employees which in no way reflect Kingspan's culture or values'.

Celotex, for its part, emphasises that a re-run of the May 2014 test without the undeclared fire barriers also passed – meaning they were not relevant to the success and failure of the test. It also stresses that all the staff members involved in the testing and marketing of the RS5000 product have left the firm. It withdrew the product from sale after Grenfell.

But lawyers for the bereaved and survivors are not so forgiving, branding them – along with Arconic – as 'little more than crooks and killers'.[30] The inquiry will make its final judgement about their actions in due course. But the facts it has unearthed so far are startling enough. For too long, Britain has tolerated a regulatory system which was, in the words of another lawyer for the community, a 'house of cards' and 'an ideal prop to facilitate industry capture'.[31] It was captured. And tonnes of combustible plastic were installed on the walls of high-rise buildings as a result.

9

2 A.M.

At 2 a.m., flames were burning the crown at the top of the building and spreading steadily but surely around it. A long arc of white flame stretched all the way up the building from Behailu's window and was edging around the north-east corner. Flames also spread in the opposite direction across the east face. Dozens of flats were on fire, smoke pouring from their windows. Thermal images from the Metropolitan Police helicopter captured what one expert would later describe as a 'waterfall of burning, molten material'. A total of 129 people remained inside the tower – almost all of them above the eighth floor.

Andrew O'Loughlin, the deputy assistant commissioner who had taken control of the incident, made his way to Command Unit 8 – the mobile van outside the tower from where the operation could be co-ordinated.

At this point – co-ordinating the information coming in from residents trapped in the tower was crucial. Residents would call the operators in Stratford and say they were trapped and give a flat number. Calls were coming in ferociously fast – in the twenty minutes from 2 a.m. they received twenty-five calls directly from the tower, as well as calls from other control rooms around the country which were taking the overflow and passing details back to London.

The call handlers still had not seen images of the fire on television. The control room was equipped with a large plasma screen television, which would normally have had rolling 24-hour news footage, but it was not working. A smaller set was not turned on. This meant they still did not have a clear picture of what was happening.

Control room staff said seeing the images on the news would have helped them understand what was happening, and may have resulted in a senior officer lifting the stay put advice. 'A picture paints a thousand words,' said one.

As it was, the advice to trapped residents continued to underplay the severity of the situation. 'The fire is actually on the fourth floor but it's obviously creating a lot of smoke,' one operator told a resident trapped on the twenty-third floor, despite the blaze at this point being at the roof and spreading around the tower. Many residents were assured firefighters would reach them in time. Particularly for those with children and disabilities, the prospect of escaping without the help of firefighters through the smoke-logged lobbies and stairwells seemed impossible. And so, they waited.

The control room would take the information about where they were and phone it through to a Command Unit at the tower. It then needed to be passed to firefighters inside the tower so a rescue crew could be dispatched.

But collating and passing on this information was proving tough. Just like at Lakanal House, the radios were overwhelmed and struggling to work inside the concrete tower. Despite their failure at Lakanal, they had not been upgraded. The messages needed to be transmitted by hand. One firefighter was asked to set up a makeshift system: standing under the south-east corner of the tower, he used the bonnet of a parked car to scribble on bits of paper the information being radioed through to him from the command unit. He then used other firefighters as runners to carry the scraps of paper to firefighters inside the tower. 'The rate they [the calls] were coming in was phenomenal. I could not write them down quick enough,' he said in his witness statement. He says at one stage there was a new call every twenty seconds.

Information was coming into the tower from this firefighter's notes, as well as notes run in directly from the command unit and radioed messages. The picture was muddled and raised the substantial risk of information slipping between the cracks.

And even when the correct information was transmitted, the job of getting to those trapped was extremely tough. The lobbies were in complete darkness with smoke and the stairs were not much better. Firefighters also had to contend with the excruciating heat on floors where flat fires were burning out of control.

The majority of the crews were equipped only with 'standard duration' breathing apparatus, which have one cylinder and give thirty-one minutes of air. This time is reduced by stress and exertion and was quite frequently not long enough to make the harrowing trip up the tower, search for a flat in the pitch-darkness, rescue the residents and bring them back down the stairs.

But some crews also carry extended duration breathing apparatus, which have two cylinders and give forty-seven minutes of air. These crews travel with larger fire engines, known as Fire Rescue Units. The first to arrive at the incident had been sent to the top of the tower by Michael Dowden to attempt to fight the fire from the roof. They never reached it. A serious number of Fire Rescue Units was not requested until ten were called for at around 2.15 a.m., and these reinforcements did not start to arrive until around 2.30 a.m. – an hour and a half into the incident. Even then, their deployment wasn't prioritised. One crew, one of the first extended duration teams to arrive, got to the incident at 1.47 a.m. but were not sent on a rescue mission until 2.44 a.m. In fact, crews with extended duration apparatus were not deployed en masse until after 3 a.m. In the meantime, firefighters were required to battle the conditions inside the tower on a single tank of air.

Faced with these conditions, the firefighters who returned to the bridgehead inside the tower, exhausted, out of air, carrying unconscious residents, were not always aware where they had ended up, or if they had reached their intended destination. They may have stopped to help a resident they found in the stairs, assist another crew who were struggling or simply gotten lost inside the building. This meant monitoring which calls had been addressed and who was still waiting was nigh-on impossible.

The rescue missions themselves were confused and frantic in the face of the blistering conditions. In one tragic incident, firefighters succeeded in rescuing a young girl from the twentieth floor. But her father and mother and two sisters were left behind. She lost her entire family.

*

On the nineteenth floor, Nick Burton was getting increasingly worried. He had gathered up the couple's passports in case they needed to identify themselves. Outside his window, he could now see the fire on the cladding.

He moved himself and Pily to the bathroom, reasoning that it was the safest place in the flat and continued to wait for the fire service to rescue them. He knew that Pily would need help to escape, considering her condition. His friend had called again – now sounding panicked – warning him to get out of the tower.

Nick redialled 999. 'We're trapped in the nineteenth floor, but nobody's coming and the flames next door are getting very close to our windows now, and the smoke's in the house,' he said.

His voice sounds calm on the call, but part of the reason for this was because he was actively trying to keep the panic out of his tone, in order to avoid scaring his wife. Smoke was beginning to fill their flat and he did not know what he would do next.

'And where do you think the fire is, next door?' the operator asked him.

'It's— no, it's— the whole tower block is on fire,' Nick replied.

'We are trying to get to you, but we've got numerous people trapped in the flats,' the operator told him.

He was told to put something against the door and to ring them back if it got any worse. The couple stayed in the bathroom. 'With water at hand and being close to the front door, it felt like the best place for us to hold out until we were rescued,' said Nick. All he could do was wait. But then he heard noises outside the front door.

*

At around 2 a.m., a team of firefighters were charged with going to a flat on the fourth floor, but when they got there it was empty. They went back to the bridgehead and were told they needed to attempt a rescue from Flat 161 on the nineteenth. This was the home of Debbie Lamprell. Despite having told the call handler she had gone up to the top floor, the crew sent to rescue her were incorrectly sent to the flat she had left at 1.30 a.m.

Being asked to go straight back up the tower was a surprise: the firefighters had already used up some of their air searching the flat on the fourth floor; they weren't sure if they would make it. The senior officer told them it was at their discretion whether they wanted to risk another mission. 'To us [saying no] wasn't an option, so we went for it,' recalls one.

As they struggled up the stairs, they were passing many residents and other firefighters coming back down. The conditions were tough. '[There was] really thick black smoke. You couldn't see your hand in front of your face. So, it was hot and you couldn't see anything,' one of the firefighters says. 'The further we got up the stairs, the worse the smoke got. As we were going up, there were firefighters coming down so we had to keep ducking into the doorways of different floors to let them past,' says the other.

As they were searching the nineteenth floor for Flat 161, they heard Nick. 'I was shouting "hello" and "we're here". I think I repeated this several times. I believe I heard a response saying something like that they knew I was there but they wanted to check the neighbours,' Nick says. He was told to wet two towels, put them over their heads and wait.

The two firefighters broke into Flat 161 and began sweeping it with a thermal imaging camera. 'I couldn't see anyone or anything. The windows were completely gone... It was very hot in there and I was conscious that we didn't have much time left,' recalls one. There were flames in one corner of the flat, and they had no water to fight the fire.

The firefighters returned to Nick and Pily's flat and Nick opened the front door. 'Immediately, the thick, black smoke billowed from the corridor outside into the entranceway to our flat. Outside it was pitch-black,' recalls Nick. An arm reached out of the darkness and grabbed Pily, another reached out for him. Realising they would not be returning to the flat, he asked about their dog, Lewis – who was also waiting by the door with a towel over his head.

'A voice told me "no" and that they were sorry but that we would not be able to take him. I looked at him. I hoped he would run out after us but he did not, he just stood there under the towel and watched and that was the last time I saw my dog, who I loved,' recalls Nick.

Assisted by the two firefighters, Nick and Pily began their descent down the dark staircase. But it was immediately obvious Pily would struggle. 'I think the lady was struggling immediately. She fell straightaway on the nineteenth floor literally as we brought her out of the flat, so we knew we were going to have a lot of trouble,' says one of the firefighters.

Progress was slow. Pily was stumbling as she went down the stairs and the firefighters were carrying her and guiding their way by holding on to the walls. Nick was also being led down by the firefighters, but he was totally blind, with the pitch-dark smoke and the towel over his head. He was reassured though by the sounds of the firefighters' voices. 'Constantly I was hearing voices shouting "let's go, keep going, let's go". The voices were firm and in charge. I remember feeling absolutely confident in their company; that they were professional; and that they would do all they could,' he recalls.

The firefighters got separated as they descended. The one with Pily was by now sometimes carrying her, sometimes pulling her backwards, slowly and steadily descending the staircase. 'It was slow progress getting down the stairs,' he recalls. 'I had to keep going down each flight to the landing or half-landing and readjusting my grip. This was repeated the whole way down the stairs. I could see on my gauge that I was very low on air and needed to get down quickly.'

About halfway down he met two other firefighters whose warning whistles were sounding on their breathing equipment. These whistles are serious – they are the final warning to a firefighter that they are low on air. Hearing them during a fire could well mean death.

Nonetheless, these firefighters joined in helping carry Pily out of the tower.

Meanwhile, Nick and the other firefighter were struggling to descend. Nick trod on things which he thought were hosepipes but now realises were likely to have been victims of the blaze. 'As we were going down the stairs I felt increasingly claustrophobic and was gasping more and more for breath. I realised that each breath I was taking in, I was inhaling more horrible smoke and I could feel my throat burning from it,' he says. The handrail became so hot that he could not hold it anymore. 'I remember thinking "bloody hell, this is it",' he says.

But the hand on his back and the voice of the firefighter encouraged him to keep going. Finally, they made it to the lobby. Nick refused to leave until he saw Pily. Eventually, he saw her being carried down by the four firefighters.

'I remember that I was just standing there gasping for breath trying to compose myself and coughing and spluttering and that when I saw Pily she was being carried by four members of the fire brigade, each on one limb,' he says.

Pily and Nick were then helped out of the building, with firefighters using riot shields to protect them from falling debris. They were out.

*

Outside the building, Tiago, the young student who had been watching Netflix when the fire broke out and was among the first to leave the building, and his family had watched the fire develop from the fourth floor and rip up the outside of the tower. Ines, his sister, was upset by what she was seeing, so she and Tiago set off for a friend's house nearby.

'All of us were just hoping the fire was on the outside,' recalls Tiago. 'I remember hearing about a fire in Dubai – I don't know what it was about but I knew it didn't spread too much inside the building. But you then started to see fire inside the building.'

Ines didn't want to look at the tower but kept asking Tiago if it had reached their family home yet. She stayed inside their friend's house and began studying for her GCSE exam the next day. 'She was just there sitting with her notes, going through them,' recalls Tiago. 'Because she didn't know what to do. At the time I didn't even consider the fact that she was going to the exam.'

Tiago's mother knelt before a statue of the Virgin Mary, praying and saying the rosary. The family were deeply concerned about some of their closest friends in the tower, Marcio Gomes, and his family – all of whom were still inside and trapped on the twenty-first floor.

*

Marcio and his partner Andreia had moved into their flat on the twenty-first floor of Grenfell Tower in 2007. Before that they'd been living in temporary accommodation in Shadwell, east London, on a waiting list for council housing back in Kensington where Marcio had grown up.

At that time, the couple already had one daughter – a two-year-old – with another on the way. Getting out of temporary housing and back to the area where all his family and friends lived was a relief.

The flat wasn't in great condition: there were holes in the walls and the wallpaper was flaking, but Marcio took two weeks off from his job as an IT manager for Ofsted to fix it up. The family were happy with their new home. There were beautiful views out across central London, stretching all the way to the London Eye and their neighbours on the twenty-first floor were welcoming, greeting them with a plate of cookies when they moved in. Over the ten years that followed, they built up strong links and friendships within the tower community. 'It was wonderful, wonderful. Everybody was so friendly,' recalls Marcio. 'I'm not claiming I knew everybody, but you saw quite a lot of people in the lifts. You had no choice really.'

Their neighbours had young children as well, and Marcio and Andreia's girls would play with them on the landing. 'Our floor, and I'm going to be biased here, was the best floor. It was just very friendly,' he says. 'Our neighbours had a young boy and he became real good friends with my youngest. We would leave the front doors open and they would play on tricycles, or kick a ball around. All those things you can do on a communal floor. It was just safe, that's one of the big things that stands out from a family perspective.'

As with most people in the tower, they had a negative experience of the refurbishment. 'I don't remember coming across any residents who wanted the refurbishment to go ahead,' he says. He welcomed the planned changes to the heating system – previously residents had been unable to control their own heating systems and the water pressure on the upper floors was sometimes too low to shower. But he rejected the need for the cosmetic changes brought about by the new cladding. 'That's just people from the outside looking at it,' he says. 'All the changes were to the outside. The inside just looked the same. All they did was wrap this cosmetic thing around it to make it look pretty. It was essentially just like putting lip gloss on it, that's all it was.'

There were also concerns about fire. Andreia worried that one of the building's two fire exits was being removed in the refurbishment and she worried that the single stair-case in the building would make an evacuation difficult in a fire. 'I remember having a conversation when we first signed the contract with the housing manager and she said if there's a fire, stay put. And I said, "Ah don't worry about that, we'll be leaving",' he says.

In June 2017, the couple were expecting another baby. Andreia was seven months pregnant with their third child, a boy they had already named Logan. 'It was just a beautiful moment for us. We never planned it, so it was a bit of a shock initially, but then that shock turned into beautiful joy,' Marcio says. 'Up until June we were just getting ready for his arrival, it was fantastic. We felt really close.' Marcio had blocked off part of a corridor and knocked through a wall in a former storage area to build a new bedroom which was already furnished with a crib, a wardrobe and a chest of drawers. The family had agreed Logan would be brought up to support Marcio's football teams: Liverpool in the UK, S.L. Benfica in Portugal. Marcio, who liked playing games on his Xbox, hoped his son would become his gaming partner. The girls were excited – looking forward to his arrival during their summer holidays from school.

On 13 June, the family had been out for dinner with some friends. They got back at around half past ten and put their daughters to bed. Andreia, heavily pregnant, went straight to sleep. Marcio stayed up to play a game on his Xbox before going to bed.

But now they were awake, and had been joined by a neighbour and her daughter. Marcio kept checking outside his front door and saw the lobby steadily filling up with smoke until it was completely dark. 'The smell of the smoke was toxic. I can only describe it as what I imagined smelt like chemicals. It was something I had never smelt before in my life,' he says. He did not feel it was possible for Andreia and the children to leave. He had been speaking to Miguel Alves, Tiago's father, on the phone and knew how the fire was spreading outside the tower. He had also made a number of calls to the emergency services – explaining that his wife was pregnant and asthmatic, that there was severe smoke on the landing and that they would be unable to leave unaided. He was told help was coming.

'Okay. I'll let the firemen know, okay, to come up to you,' said the call handler in a call which began at 2.19 a.m.

'Three, three kids and three adults and one heavily pregnant wife,' replied Marcio.

All they could do was wait. Smoke was beginning to seep into their flat around the edges of the door. Time was running out.

*

The communication of information from the control room in Stratford to the incident ground and then into the tower was branded 'a deplorable state of affairs' in the final judgement of Sir Martin Moore-Bick.[1]

Sir Martin said that what was needed were single, clear lines of communication and an overall command structure. But this was lacking. The LFB had failed to properly plan for an event with mass calls from trapped residents and therefore the senior officers had no plan to put in place – despite the devastating experience of the Lakanal House fire only eight years previously. The 'information loop' between the bridgehead in the tower, the command units outside and the control room in Stratford was, he said, 'never completed'. No one told the control room the fire had spread out of control, which meant they kept telling callers that the fire was on the fourth floor when it was not.

This had severe consequences. A family of five on the seventeenth floor first phoned 999 at 1.29 a.m. and spoke to the police, who passed on their location to the LFB's control room in Stratford. Their details were passed to the command unit at 1.43 a.m., but did not make it onto any of the pieces of paper carried into the tower. The message did seem to have been relayed to the bridgehead inside the tower, but no deployment was made. During the 999 calls, the family were told repeatedly and unambiguously that help was coming and that the safest option was to stay put. Their youngest son was only eight, the landing was full of smoke and it would have felt extraordinarily dangerous to flee unaided into the dark, choking smoke outside. The family phoned 999 again at 2.27 a.m., warning that the fire was right outside their window and they were afraid they would die. This message was radioed over to the command unit outside the tower, but it is not clear what happened to it. Further calls were made at 3.09 a.m. and 3.18 a.m., and the family were told to 'make a run for it' but could not leave due to the smoke. 'I could have got out a long time ago, we could have, but they said, "Stay in the flat, stay in the flat,"' said the father in this call. The family was never reached. All five of them died. It is estimated they could have fled up until 2.45 a.m. and survived.

Below them, on the fourteenth floor, a similarly tragic story was playing out. There, eight trapped residents – including a two-year-old baby and his mother – were reached by firefighters at 1.51 a.m. But rather than lead them down the stairs, the firefighters moved them into Flat 113 – the least smoke affected flat on the floor. Despite the pleas of some of the residents to take them out of the tower, the firefighters reasoned that it was safer to return with more help. But no one returned until after 2.30 a.m. when two crews reached the fourteenth floor. These firefighters had limited and incorrect information about who was trapped and where. In the dark and confusion, they led four of the eight residents to safety and left the other four behind. These residents were never reached, and perished – the last to succumb dying after 4 a.m. Despite further calls for help from this flat, crews with extended duration breathing apparatus initially sent to help them were 'inexplicably' redeployed to fight fires or search lower floors. The two-year-old boy died crying and coughing in his mother's arms while she spoke to a firefighter outside the tower on the phone. She lost consciousness moments later and also died at around 3.45 a.m.

In fact, only three phone calls from trapped residents resulted in a fully successful rescue mission, and partially successful rescues from two further flats. Everyone else either escaped unaided, came across a crew sent to another rescue like Pily and Nick, or never made it out of the tower at all.

10

'WE WILL BE QUIDS IN!'

It is now clear how dangerous cladding and insulation materials became widely available in the UK: the bitter fruit of corporations capitalising on deregulation. We now need to turn to the story of how they ended up on the walls of Grenfell Tower itself.

The sorry state of the construction industry is unveiled in this saga: costs were cut, materials switched for cheaper options, quality of workmanship was desperately poor and oversight all but absent. These failures are systemic in an industry in dire need of reform. It is not just a story of one shoddy construction job, but a deeper one concerning the neglect of social housing and the degradation of the entire construction industry. To tell it, we must return to the beginning.

THE ROOTS OF THE REFURBISHMENT

Grenfell Tower was built in 1974. By the 1970s, the surrounding area of North Kensington was diverse, having welcomed a community of Spanish refugees fleeing the

Civil War and the Windrush generation of immigrants from the Caribbean. The housing conditions in the area were poor: old Victorian terraces had fallen into the hands of exploitative rogue landlords. Branded 'slums' they were bulldozed and replaced with a modern housing estate: the 795-home Lancaster West. This construction project completed in 1974, with Grenfell Tower at the heart of the new estate.

Those displaced through the demolitions were rehoused in homes built to spacious design standards. 'My neighbours who moved in at the time have told me how incredible it was,' says David O'Connell, a current resident.[1] 'They were going from houses with five people per room to a nicely sized family home.' But this bright promise soon faded. Plans to include offices, shops and amenities were never realised.[2] Shortly after the estate's completion, the cross-party political consensus about building and investing in state-owned housing died with the election of Margaret Thatcher in 1979.

Thatcher believed housing should be provided by the market, not the state. Her reforms choked the power of councils to invest in the maintenance of the hundreds of thousands of council homes built by Tory and Labour governments in prior decades. Their condition started to deteriorate. At the same time, the Right to Buy was introduced – allowing council tenants to buy their homes at a discount. Homes were sold to the tenants and sold again at a profit. Many became the property of private landlords, let out at far higher rents than the council would have charged.

This happened across the country, but it was particularly noticeable in North Kensington, where residents of the council estates, which were being allowed to fall into disrepair, lived just streets away from some of the wealthiest, most exclusive addresses in the country.

Local politics in this part of London has always been fraught. The council has been run since its formation by councillors elected by the rich residents. But it is responsible for the lives of the poor, who are more reliant on its services. This is particularly true for social housing, where it is not only the community's political leader but its landlord. There have always been accusations that the council does not care about the social housing it manages and spends too little time and money on its upkeep.

By 2010, residents of the Lancaster West estate were fed up. They felt the estate was being mismanaged and their concerns were going unaddressed. The estate had received no major investment in the thirty-six years since it was built. Around Kensington and the rest of London other estates which were becoming similarly rundown began to be listed for 'regeneration': demolition and replacement with private housing. The council's rules did not guarantee them a right to return to the estate if it was demolished, and they feared being scattered across London and beyond to make way for richer residents to move in. '[The council] didn't invest in [the estate] at all in the years before Grenfell,' says Abbas Dadou, current chair of the residents' association. 'We always worried their plan was to knock it down.'[3]

Plans were under consideration for such a regeneration at the Lancaster West estate. A draft 'masterplan' for the area, produced by a private consultancy for the council in 2009 recommended demolishing 'most of the existing housing in the area' to make way for new build. It noted that interest in the project from private developers would be 'significant'. This report specifically singled out Grenfell Tower, saying its 'appearance...blights much of the area'.[4] One of the directors of the consultancy which created it was hired by the council and made executive director of planning in 2010.[5]

'I believe the council recognised that the land we were living on was a goldmine and they didn't even have to dig for the gold,' Eddie Daffarn said in his witness statement to the inquiry. 'All they had to do was marginalise and displace the people that were living there, and by doing so they could then replace the social housing with "mixed communities", make a fortune from property develop-ment and sales [and] change the electorate in those wards.'

The council certainly did believe in estate 'regenera-tion'. Rock Hugo Basil Feilding-Mellen was made deputy leader of the Royal Borough of Kensington and Chelsea and cabinet member for housing in 2011, aged just thirty-two. He was a passionate advocate of the demolition and rebuild of council housing estates. A rising star in the Conservative Party, Mr Feilding-Mellen is a descendant of British aristocrats whose family home is Beckley Park, a Tudor hunting lodge with three towers and three moats. It is quite possible that he would have led RBKC into a regeneration of North Kensington, and the Lancaster West estate and Grenfell Tower would have been demolished along with dozens of other council estates across London.

But times were changing. US banks had gambled too hard on the housing market and in 2009 the world plunged into a shattering recession. The appetite from private developers to rebuild the estate and replace it with luxury flats disappeared. The regeneration plans were canned. But the estate would not simply be left alone. Instead, the council planned to build a leisure centre and academy on the ground outside the tower.

For many residents, this project – the Kensington Academy and Leisure Centre, or KALC – was the final straw. It entailed a lengthy construction project on their doorstep and would have removed the only green space they had, the only outside space they had to walk their dogs and let their children play. 'Naturally we were going to have concerns,' recalls Eddie Daffarn. 'A piece of land surrounding our estate that had basically been subject to managed decline was suddenly going to be taken by the council. But we weren't listened to, we weren't treated with any understanding.'

As we have seen, Mr Daffarn and neighbour Francis O'Connor launched a WordPress blog (Grenfell Action Group) to campaign against this and other issues, including the safety of the tower. One of the points raised by the campaigners was that the expenditure on KALC ($£17m$) was unfair when nothing had been spent upgrading the tower.

Partly to quieten these concerns, the council decided to carry out refurbishment work to the tower alongside the construction of the academy and leisure centre. There were some serious problems with the building by this point: the communal boiler needed replacing, windows were old and draughty, lifts regularly broke down and the building was expensive to heat. But this was not the council's only concern. It also worried about the appearance of the tower, which it felt could end up 'looking like a poor cousin to the brand-new facility being developed next door' according to a 2011 email.[6] It planned a refurbishment: the communal boiler was to be replaced, community and office space at the base of the tower would be converted into new flats, new windows would be installed. And to address concerns about the tower's thermal performance and appearance, an external cladding system would be added to its walls.

Funding for the work was cobbled together from the sale of some basements the council owned near the King's Road – one of the most exclusive addresses in the UK. This raised £6m, which was topped up to £8.5m with the injection of some council funds. For the amount of work required, this budget looked low.

To some extent, this was a local political decision. RBKC prided itself on achieving 'value for money' and was determined to ensure it got the lowest possible price for the work. But on another level, it reflected national austerity. While RBKC had substantial cash reserves (£274m), these were on the 'general fund' side of its business, and local government rules prevented them being invested in housing. Why didn't it borrow the money to properly fund the refurbishment? Local authorities could borrow at exceptionally low rates at this point in time, and repay over several decades through the rents from their housing stock. But the government had banned councils from taking this step. Obsessed with bringing down national debt – inflated as it was by the decision to bail out the failed banks – the Treasury had imposed a cap on council borrowing for the simple reason that it would appear on the national balance sheet. The refurbishment would be done on a shoestring. Costs would be squeezed to make it fit.

2012

In a bid to achieve what was described as 'economies of scale', the council sought to appoint the same contractor (Leadbitter) and architect (Studio E) who were working on the new school and leisure centre to refurbish the tower as well.

This was an odd decision. Studio E – which specialised in education – had no experience of this sort of work. Minutes of a 2012 meeting with residents record Eddie Daffarn asking if the architects 'have experience with tower blocks and if not why are we using them?'

This is a question that should have been asked formally. Procurement rules meant the job should have been advertised publicly and given to the winning firm. With its lack of experience, it is very unlikely Studio E would have won this bid. But this work simply did not happen. Instead, the firm's fee was split in half to avoid the threshold which required a procurement process and the architects were appointed without scrutiny.[7]

Architect Bruce Sounes was put in charge. It was to be his first high-rise and his first residential cladding project.

In 2012, he set about drawing up plans for the tower's new cladding system. But, according to his own testimony, he did not check the relevant regulations and rules. His view, he explained to the inquiry, was that cladding specialists would be joining the project later on. It would be up to them, he felt, to spot and correct any mistakes.

To recap – at this point Approved Document B required Mr Sounes to choose a Class 0 cladding panel and limited combustibility insulation (in practice, this meant mineral wool), or to use a system which had passed a large-scale test. But alongside consultants Max Fordham, the architects opted to pursue an 'aspirational' thermal performance for the tower. They wanted to insulate it to the same standard as a new build. As a result, Mr Sounes and the consultants decided Celotex insulation was the only material that would achieve the required performance.* It did not meet the minimum standard of limited combustibility and had not, at this point, passed a test. But Mr Sounes was aware it was widely used in the industry and assumed that it was compliant.

For the cladding, Mr Sounes selected zinc panels. He did not consider their fire performance, but liked the way they looked. Residents were consulted in August 2012 and were told they would be getting 'fire-retardant' cladding.[8]

It is worth making clear that there were plenty of fire safe options for this job. As well as Mr Sounes' zinc panels,

---

* Inquiry transcript, 24 September 2020. An expert witness at the inquiry has queried this, demonstrating a design that could have met the required performance with non-combustible insulation.

D+B Facades was asked to cost the provision of solid aluminium sheets for the project which would have been fully non-combustible.[9] I have also spoken to a supplier of rendered systems – which would have involved applying non-combustible insulation directly to the walls. But this does not provide the sharp, modern look of a rainscreen system. 'They were favouring the rainscreen cladding system for aesthetic reasons – the thinking was "we can make these concrete buildings look like shiny new buildings". The rainscreen cladding was preferred because it delivered a sharp, contemporary aesthetic,' the supplier said.[10]

While Mr Sounes had little knowledge of fire regulations, he had the support of consultants who did. Exova, one of the world's leading fire safety consultancies, were engaged to produce a report into the early refurbishment plans. Here was an opportunity to warn about the potential fire safety dangers of cladding systems and point out that the combustible insulation being specified was not compliant.

But Terry Ashton, the consultant appointed to produce the report, was not a qualified fire engineer. His experience came from years spent working in building control, signing off projects as compliant. In an early review of the plans for the refurbishment, one of his Exova colleagues had said in an internal email that the work would 'make an existing crap condition worse' in the tower.* She asked if the

---

* Inquiry transcript, 16 March 2020. The 'crap condition' was a reference to escape distances in the newly added flats, not the cladding.

firm knew anyone in the building control department at RBKC, which would ultimately sign the work off. 'Terry's the man for contacts down there,' a colleague replied – in reference to the fact that Mr Ashton had worked with two of RBKC's inspectors earlier in his career.*

Mr Ashton did not raise issues with the cladding. In October 2012, he drafted his outline fire strategy for the refurbishment. In a list of major changes, his report did not even mention cladding was planned – focusing instead on the plans to convert the lower floors from commercial to residential space. His section on external fire spread simply recorded: 'It is considered that the proposed changes will have no adverse effect on the building in relation to external fire spread, but this will be confirmed by an analysis in a future issue of this report.'[11] He had, in fact, been sent plans for the cladding, which included the non-compliant insulation which had been selected. But he had not even opened them. He said his belief was that the question of cladding would be dealt with at a later date, at which point he would give a more informed opinion.[12]

And so by 2012, the project had already started down the road to disaster. But they had hit an early problem with finance. Leadbitter had quoted between £12m and £13m for the job, but the council's budget was limited to £8.5m. The job stalled, and cost savings were sought before it could get going again. These savings would lead to Grenfell's demise.

---

* Mr Ashton insisted this gave him no advantage in seeking sign off – saying that former colleagues may even have been tougher with him.

2013

In early 2013, the planned project was at a virtual standstill due to a disagreement about cost. It came close to being abandoned altogether.

The project was being run by the Kensington and Chelsea Tenant Management Organisation (the TMO), the arm's-length organisation set up by RBKC in 1996 to manage its 9,000 council homes. Management consultants Artelia told the TMO it should back away from the project. 'Unless the project, in its current guise, is stopped and a review embarked upon to redefine the scope, programme and cost, it [the refurbishment] will fail,' the report said.[13]

But this advice was not taken. A new director, Peter Maddison, had recently joined the TMO and was responsible for overseeing the refurbishment. He suggested trying to remove Leadbitter from the job and find another contractor willing to do it more cheaply. 'Value for money is now the primary driver for the project,' notes from a project meeting in spring 2013 show.[14] Artelia was asked to produce a new report giving new advice which supported this course of action. The objective was now getting costs down to RBKC's budget.

The phrase 'value engineering' means reducing the cost of a construction project without reducing its quality by finding new materials or cheaper techniques. In reality, costs are cut by substitution with inferior alternatives. With costs perennially pushed down and profit margins always tight, this happens across the construction industry every day.

As early as February, Mr Sounes identified the potential to switch 'zinc cladding to something cheaper' to strip cost out of the job. As well as speaking to Debbie French from Arconic, he had also been contacted by Harley Facades which had found the plans for the refurbishment online and were angling to win the job. On 27 September, Mr Sounes had a coffee with them in central London.

At the meeting, they told him his preferred zinc cladding would cost around three million pounds. But there was an alternative, they said, showing him glossy brochures of other tower blocks Harley had clad with a cheaper product: ACM. 'They said their recurring experience is that budgets force clients to adopt the cheapest cladding option: ACM,' Mr Sounes wrote in an email shortly after the meeting.[15]

Mr Sounes continued to exchange emails with them – asking for quotes on various zinc products. They continued to emphasise ACM. 'From a Harley selfish point of view our preference would be to use ACM. It's tried and tested on many Harley projects,' wrote commercial manager Mark Harris in one email.*

Along with a colleague at Studio E, Mr Sounes was now required to draft a specification for a new contractor to bid for the job. For the cladding panels, Mr Sounes was still convinced that zinc was the right choice, because

---

* Inquiry transcript, 9 September 2020. Mr Harris described this as a 'poor choice of words', adding that he simply felt ACM was a good product and may provide a good solution for Grenfell Tower.

of aesthetics. But he was also required to offer alternative products. Here, he listed Reynobond ACM made by Arconic. The seeds of disaster were sown.

2014

Public procurement is done to ensure contracts funded by public money go to the best candidate. This is an important part of the democratic process: a guard against the nepotism, waste and corruption that blights government projects across so much of the world, and to which the UK is in no sense immune.

But it is also a process that can drive an unhealthy focus on cost reduction. In order to win the job, potential contractors have to submit bids. They know that they are overwhelmingly likely to win if they enter the lowest price. This encourages them to fill their order book by bidding low, and then solve the problem of how to squeeze a profit out of the job later down the line.

By early 2014, three companies had entered formal bids – Durkan at £9.9m, Mullaley at £10.4m and Rydon at £9.2m. This made Rydon the frontrunner, but still put it £700,000 above the TMO's budget. Peter Maddison, who had several contacts at Rydon from previous jobs in housing, got in touch informally to see if the firm could strip a little bit of extra cost out of its bid – something which went against the strict rules of transparency which are supposed to govern the procurement process. On 10 March, Rydon director Steve Blake wrote in an email:

'Spoke to Peter [Maddison] about the award and they are keen to get going. They need to do a fair amount of value engineering which should be achievable.'[16]

On 18 March, representatives from Rydon and the TMO sat down to discuss in detail the cost savings they could make. The meeting was secret (described as 'offline' in internal emails) and no official notes were taken. The two firms left content that the £700,000 savings they wanted could be made. Later the same day, Rydon was formally appointed. But the agreement would involve switching out the zinc panels for the alternative: ACM.[17]

It would later emerge, at the inquiry, that Rydon did not tell the TMO the full extent of the money that could be saved by making this switch. A project manager accepted its intention was to 'pocket the difference' to boost its profit margin. 'We will be quid's in!' said one staff member in an email.[18]

The cladding switch required sign off from planners, who were shown the certificate from Arconic's saleswoman Debbie French which suggested the panel was Class 0. That was enough. In fact, the only serious questions asked about the cladding related to aesthetics.

The planning committee insisted on cassettes which they felt would look smoother and attract less rust, not knowing that these were the most violently combustible. And Rock Feilding-Mellen chipped in to express his views on the colour: arguing about options such as 'champagne' or 'British Racing Green'. Fire risk did not come up.

It now fell to Harley – now appointed as a subcontractor by Rydon – to work out the specific construction drawings for the cladding. But no one took overall responsibility for making sure what was being planned was compliant with the law or safe. Mr Sounes said he expected Harley to check his designs were suitable. Harley's witnesses said they assumed Mr Sounes had already done these checks. Rydon's witnesses said they relied on their subcontractors to get it right, and the fact that final sign off would be offered by building control. Everyone looked to someone else to ask the key question about safety.

What about the fire engineers, Exova? Terry Ashton had updated his report in autumn 2013, but did not change the line which said the project would have 'no adverse effect' in relation to external fire spread. His report still said this was to be 'confirmed' in a later version of the report. But this final version was unlikely to be forthcoming. Amid its efforts to reduce the cost of the project, Rydon had stopped paying Exova to advise on the project.[19]

Nonetheless, other members of the project team did not know this and continued to email Mr Ashton for advice. In September 2014, he was asked a technical question about cavity barriers by one of the team at Studio E. His response suggested it was acceptable to use combustible insulation if cavity barriers were provided. This was wrong. Combustible insulation should not have been used on a high-rise at all – except in the very specific instance of a system which had passed a large-scale test. The email was described by an expert witness to the inquiry as 'incorrect and inaccurate'.*

With no one in the refurbishment team properly checking the compliance of their plans, the burden fell to the council's building control inspector to spot the flaws.

But the building control department at RBKC was under pressure. Since the 1980s, builders have been able to appoint private inspectors to sign off projects. This has put council services under pressure: the private sector pay better and recruiting and retaining staff at town halls is difficult. But despite this, local authority inspectors are obliged by law to take on all applications they receive. Inspectors at councils became overworked.

And then austerity hit in 2010, and councils began to streamline and cut building control departments and back office staff. Teams were cut to the bone. At RBKC the team was restructured to make it 'self-funding'. The number of staff employed by the team fell from twelve inspectors to five. Research I carried out in 2021 suggests this was by no means unusual: the number of inspectors employed by councils had been slashed by 27.4% across the preceding decade, with one council dropping from ten to just two.[20]

---

* Inquiry transcript, 28 October 2020. Mr Ashton argued when questioned that it was a quick email written without full consideration. because he was no longer a fully paid up member of the design team and he wasn't being asked to give an overall view on the plans for the cladding system.

At RBKC, the council had previously had a 'special projects' team of senior, specialist inspectors who would take on the most complex jobs and ensure they were handled properly. But this team was disbanded in the restructure and the borough was divided into geographical patches. If a job fell in a certain inspector's postcode, it was theirs – regardless of their skill and experience.

Grenfell Tower fell onto the patch of John Hoban. He had worked at the council for twenty-seven years. But he was not part of the special projects team and had focused on simpler jobs as a result. Nearly three decades into his career, Grenfell would be his first special project and his first cladding job. He was also under intense pressure: working evenings and weekends to stay on top of his workload.

He approved Rydon's plans in September 2014. The cladding panels were still labelled as zinc, even though the decision had already been made to switch to ACM. But Mr Hoban would never request a comprehensive description of what would be included in the cladding system. Mr Hoban said he 'drew comfort' from the fact that Exova were advising on the project – assuming their advice would ensure the plans were compliant with fire safety rules.

And so as 2014 drew to a close, no one was asking questions about fire safety. Except once. On 12 November 2014, Claire Williams, project manager at the TMO, was reminded of the Lakanal House fire and struck by a fear that the Grenfell Tower project was also making use of dangerous materials.

'I have just been looking at the cladding as our database is asking for costs… However, I do not know if there is any issue of flame retardance requirement? I know at Lacknall [sic] House one issue was that the replacement panelling for the asbestos cladding was not flame retardant! I don't know if this is in the specification, but want to make sure it is raised. Please advise,' she wrote.

Artelia told her to contact Rydon. She did, sending an email to their project manager saying that she had just had a 'Lacknall moment' and asking for clarity over the fire retardance of the cladding. There is no evidence of a response and no evidence that she or Artelia followed it up. This would later be described by lawyers for the survivors as the last chance to avert disaster.[*]

2015

Installation of the cladding system was ready to begin in early 2015. Three thousand square metres of cladding panels were sold by Arconic to CEP, a fabrication firm, in

---

[*] Inquiry transcript, 19 October 2020. Asked about this email exchange at the inquiry, Mr Lawrence claimed he read it as only relating to the glass reinforced concrete cladding on the very ground floor of the tower. This is because a reference was made to it in Ms Williams's email. He also said it would not have been something he had the skill to answer. Ms Williams could not recall what happened, but suggested Mr Lawrence may have spoken to her in person to offer a reassurance. She claimed she was given separate assurances by Mr Lawrence that the cladding was 'inert', something he denies.

March 2015. They were bent into the cassette shape and delivered to the site. No one would ask about fire, and no one would be told a more fire-retardant version of the panels was available for a mere £2/£3 per square metre on top of the £23 per square metre cost. The cost of using a safer panel would have been less than £9,000 – or 0.1% of a job now worth more than £9m.[21]

Inspector John Hoban learned ACM was being used at some point in 2015, but he could not say specifically when. He told the inquiry he 'just looked at the first page' of the BBA certificate and accepted it 'without question'.[22]

Mr Hoban learned the Celotex insulation was being used when he saw the boards being delivered to site. He looked them up online and found the Local Authority Building Control website which implied the insulation was suitable for use on high-rise buildings.[23]

While Mr Hoban did not raise queries about the use of the combustible materials, he did, in March 2015, raise some questions about the planned use of cavity barriers – which are supposed to prevent fire shooting up the gap between the insulation and the cladding panels. This was a technical debate: he believed that instead of simple barriers the system required a tougher form of fire break known as 'fire stopping', which would have two hours of fire resistance. But the design team felt this was unnecessary. On 27 March 2015, Daniel Anketell-Jones of Harley wrote: 'It's ridiculous. There is no point in "fire stopping", as we all know; the ACM will be gone rather quickly in a fire!'[24]

This startling email triggered headlines around the world when it was first revealed by the inquiry in January 2020. It may not be as shocking as it first appears: Mr Anketell-Jones and others were at pains to stress that the risk he was referring to was a panel coming loose and falling off the building during a fire – not igniting and spreading fire up the building. But it does show the team was aware of safety risks of some kind. And while ACM was widely used in the UK, its fire performance was no secret by 2015. Fires in the UAE had been reported in the industry press and on the BBC. Warnings about the product were a mere Google search away. Mr Anketell-Jones insisted his knowledge was limited to structural matters and he knew nothing of the fire risk of the materials being used. But at the inquiry, he was asked about emails he sent which included detailed points about fire performance, and a conference he attended, where academics gave a presentation that focused on the mechanisms of fire spread on cladded buildings and included a series of case studies on major cladding fires in the UK and abroad. He said he could not remember it and may have been out of the room at the time.

There was also a problem with the cavity barriers which were due to be installed around the windows. The regulations required a barrier right above the window opening, to prevent fire spreading out of the window and up the building. This was noted by the cavity barrier supplier who circled a design drawing and annotated it with the words 'weak link for fire'. Ben Bailey, the 25-year-old son of the Harley Facades' director who had been made project manager, didn't do anything to follow up the warnings.*

By June 2015, work to install the cladding system was well underway. The project was going badly: deadlines were missed and budgets were stretched. 'At the moment we have a poor-performing site which is mainly but not totally caused by poor surveying and cheap incompetent subcontractors,' wrote Rydon's project manager in an email to his boss.**

Osborne Berry was the building firm appointed by Harley to actually install the cladding. Post-fire investigations would reveal serious problems with their work. Cavity barriers were installed upside down, vertically when they should have been horizontal, in the wrong place and cut to the wrong shape. No vertical cavity barriers were found on the tower at all. A member of staff at the supplier of the barriers would later call it some of the worst work he'd ever seen.[25]

This was despite the work being checked by numerous professionals who went up and down the tower to review the cladding as it was being installed. But the inspections were largely cosmetic and focused on minor details: scratches and scuffs to the cladding, gaps between the insulation panels which needed to be taped. A clerk of

---

* He said he believed the warning was given in the specific context of the debate about fire stopping, which was resolved when Mr Hoban accepted cavity barriers were acceptable.

** He said this was just him 'having a vent'.

works, employed by the TMO to ensure the quality of the project, made ten trips up and down the tower to check the cladding. He later told the inquiry that his impression of the cladding works overall was that it 'looked very neat'. John Hoban also inspected the project on an almost monthly basis. But he only inspected a small percentage of the cavity barriers – many were installed and immediately covered up by the external cladding panels before he saw them.[26]

It is fair to say that the experience of the refurbishment was miserable for the residents of the tower. Women with new-born babies were stuck in flats with the noise of drilling and banging outside as contractors fitted the cladding rails to the walls of the building. The loose cladding rails banged against the walls during the night. With workmen using the lifts, there were often long queues – people were late to work and school in the morning and those with mobility issues struggled to get in and out of the tower. An email referred to workmen deliberately winding up residents: banging on windows as they travelled up and down the tower to scare pets or ask for cups of tea.*

But what really riled the residents was the decision during the job to move the position of the units which delivered hot water from the communal boiler to their flats. This was done to speed up the work, but it was extraordinarily inconvenient. It would have forced those

---

* Inquiry transcript, 29 September 2020. The contractor alleged to have done this denied it.

coming in and out of the flat to squeeze past a big, bulky boiler which had sharp edges at roughly the height of a child's head. Many residents refused. They claim they were told they would lose their hot water if they didn't have it done, and that workmen would appear when elderly residents were home alone and do the work without the consent of others. Relationships between residents, the TMO and Rydon – already miserable – plummeted to an all-time low. Signs started going up on doors around the tower saying 'no access' for Rydon workers.

By now, the TMO had effectively shut down consultation on the project, fed up of the complaints being raised by Eddie Daffarn and other dissatisfied residents. Internal minutes from 2013 record that 'it has been agreed to hold no more public meetings due to the stand being made by the Grenfell Tower leaseholder group.'* As the row over the heating units intensified, Mr Daffarn and a neighbour, David Collins, sought to set up a group to collectively represent the residents of the tower, branded 'Grenfell Community Unite'.

But the TMO refused to recognise the new group. Internal TMO emails branded the planned group 'a showcase for Mr Daffarn' and said they would not meet with them. 'Mr Daffarn is continuing to agitate in Grenfell Tower,' wrote Peter Maddison in an internal email, asking

---

* Inquiry transcript, 4 May 2021. Asked about this at the inquiry, Peter Maddison, director of assets at the TMO, said it had been 'inaccurately recorded' and the in-person meetings were wrapped up because they were poorly attended and inefficient.

one of his colleagues to investigate whether his comments had 'become libellous'.[27] In the end, it took the intervention of the local Conservative MP to bring the TMO to the table with Grenfell Community Unite. The complaints about the position of the heating units were raised and Rydon and the TMO agreed to revert to the original plan. The power of collective resident action was powerfully demonstrated.

Nonetheless, the meetings did not resolve all of the tension. Footage later obtained by Channel 4 showed the MP, Victoria Borwick, telling a mother who was complaining that her water had been cut off by the work to 'take baths with people next door'.[28] As resident complaints about the refurbishment mounted, Rock Feilding-Mellen would exchange emails with the council leader, Nicholas Paget-Brown, where the latter described the complaints as a 'dry run' for the sort of concerns 'that will come up once we get nearer to proposing some actual estate renewal'.[29]

The anger about the project did not die down. In December 2015, with the support of local councillor Judith Blakeman, Mr Daffarn gave a speech to a full council meeting outlining what he believed had gone wrong and presenting a petition of residents angry about the condition of the tower.

'There are far too many examples of poor quality works and poor site management to list to you individually, but some examples include a number of residents who were left without any hot water for months on end, a vulnerable, disabled, non-English-speaking woman being denied a toilet in her own home for three days and being forced to walk to a friend's in Shepherd's Bush in order to take a shower for three months,' he said.[30] The residents wanted an independent review to assure the safety of the works that had been carried out. They would not get it.

2016

The council did agree to a review of the project, but said this would be carried out by the TMO itself. The organisation was being asked to mark its own homework. Why was there not more scrutiny? An oversight committee within the council, which was supposed to keep a check on the TMO's work, does not appear to have been particularly concerned by residents' complaints. In March 2016, the chair of this scrutiny group emailed the local MP to say: 'This [the refurbishment] is essentially a £100k gift from the state. I'm therefore not massively sympathetic to general "it's all terrible" complaints.' He added that the residents' complaints were 'grossly exaggerated' and 'should be taken with a large pinch of salt'.[31]

The TMO's investigation took the form of a review lasting 'one full day'. It involved a presentation about the refurbishment, a tour of the tower and an information pack covering the issues raised by the tenants. In the event, the TMO produced a six-page report which suggested small tweaks to the consultation process and claimed Rydon had 'responded adequately' to complaints – which it said had been submitted by just four residents. It concluded by 'commending the contractor Rydon on [its] performance' and the TMO for its 'high quality management of the project over months'. The chance to ask difficult questions about the safety of the refurbishment was lost.[32]

Finally in May 2016, over time and over budget, the refurbishment was complete. 'It is remarkable to see first-hand how the cladding has lifted the external appearance of the tower,' said Nicholas Paget-Brown in the council's official press release.[33]

John Hoban issued a completion certificate in July 2016. He did not have legally required safety information, but signed off the project anyway. He had not satisfied himself that the cladding system complied with Approved Document B. Months later, he would resign. The pressure had become too much and his health was being affected.[34]

The residents – frustrated and worn out by the years of construction work – were glad that it was over and they could return to peaceful enjoyment of their homes. But they worried about the standard of the work. The windows were draughty, as if there were gaps in them, and their homes were colder as a result, despite the promise of better insulation. They witnessed builders who had been called back to fix the defects fill gaps up with plastic foam. A few wondered what would happen to all this plastic in the event of a fire.

Eddie had not given up raising concern. In August, he saw reports of the fire in Shepherd's Bush, which spread over five floors. In November 2016, he uploaded his blog post titled 'Playing with fire' and predicted 'an incident that results in serious loss of life'. He did not know how quickly his prophecy would come true.

II

2.30 A.M.

Shortly before 2.30 a.m., one column of flame edged from the east face of the building around to the south. On the north face of the building fire now stretched in a diagonal line from the bottom corner to the top. Behind it were wrecked flats and black, burned out cladding and insulation. In front of it were the flats which were still unaffected, and more shiny panels full of fuel.

Inside the building, the lobbies on all levels of the tower were now at zero visibility, filled with dense, irritant, toxic black smoke. The stairwell was still just about passable, but increasingly full of smoke. Residents' flats were also becoming increasingly smoky. This meant they were inhaling increasingly dangerous amounts of toxic chemicals which reduced their chances of leaving the tower alive by the minute.

This smoke was a horrible mix of cyanide, carbon monoxide, nitrogen oxide and irritant acid gases from the burning insulation and fire-retardant chemicals in the materials used to make the window surrounds. The smoke on the landings was also full of the products of burning furniture – which were stuffed full of chemical flame retardant. All of those who remained in the tower were gradually building up higher concentrations of these chemicals in their bodies, to a point where they would eventually collapse. Collapse occurs when the percentage of carbon monoxide in the blood tips over 20%. This happens suddenly. Those whose carbon monoxide levels had reached 17% or 18% while they waited for rescue in their flats entered the lobbies or stairs, breathed in the highly concentrated smoke, collapsed instantly and continued inhaling smoke until they died. Others, who had inhaled less smoke while they waited, could potentially survive for fifteen minutes in the stairs before collapsing. Survival for those still trapped in the tower was becoming less likely by the minute.

By this point, 177 of the 293 occupants of the tower had left. Many had run out of the tower in the very early stages, through relatively clear lobbies and down the stairwell before it filled with smoke. Approximately half the occupants of the tower had made it out this way before 1.30 a.m. without serious exposure to toxic smoke. The rest had managed to battle through increasingly difficult conditions in the lobbies and the stairs in the hour since. But by 2.35 a.m., 116 people were still inside Grenfell Tower.

*

Andy Roe was the second most senior firefighter in the London Fire Brigade on the night of the blaze. He arrived at 2.31 a.m. and as he walked towards the building he saw it burning from the third floor up.

'I began to encounter small groups of very agitated friends and family of those in the tower. More than one group told me their relatives or friends were trapped and were on the phone to them,' he recalled in a witness statement. 'As I got closer to the base of the tower it was clear small groups of Police were preventing members of public entering the tower but they were in danger of being overwhelmed by the numbers.'

He reached the command unit outside the tower and received a quick briefing on the situation from Andrew O'Loughlin. He was concerned about the advice to stay put. He felt the policy had become 'absolutely unsustainable'. The building had failed to such an extent that anyone who was still above fourth floor was 'in great danger' and needed to be told to take their chances getting out – whatever the risks on the stairs. He told one of the officers in the command unit to phone the control room in Stratford and inform them of the decision to tell residents to leave.

As it happened, officers on the other side of the city in Stratford were coming to the same conclusion. Joanne Smith, a senior operations manager, had arrived in the hectic control room in Stratford at around 2.15 a.m.

She began by listening in to two calls from trapped residents. What she heard caused her to worry her that the stay put policy was no longer working. 'Both calls indicated that the situation was getting worse in terms of smoke and heat,' she said in her evidence to the inquiry. 'I became increasingly uncomfortable with the stay put policy.'

The television in the control room was still turned off. But at around the same time as Joanne Smith was having these thoughts, a senior firefighter who was present in the room walked downstairs and saw an image of the fire being broadcast on Sky News. It was the first time he had seen an image of the tower, with its east side fully engulfed in flame. He rushed upstairs to tell Ms Smith what he had seen.

That was enough for the decision to be made. She instructed call handlers to change the advice: residents would now be told to leave when they phoned the control room. One of the operators scribbled the new advice onto a piece of paper and held it up to the room: 'Tell callers to put wet towels over their heads, cover their faces, hold hands, and leave.'

But such was the embedded faith in stay put among call handlers that this change in policy was not followed immediately. Between 2.36 a.m. and 2.42 a.m. – immediately after the advice changed in the control room – five separate call handlers took six 999 calls between them and in no case did they advise the caller to leave. Instead, the advice remained to stay in the flat or wait for rescue.

But, finally, the control room and incident ground agreed – residents needed to get out of the building. The decision to lift the stay put advice completely was recorded in the LFB's log of the incident at 2.47 a.m., just under two hours from the first 999 call and nearly eighty minutes after flames reached the roof of the building. For many it was already too late.

*

Why did it take so long to lift the stay put advice? We will come on to the complex position of government guidance and how this country came to place such enormous faith in the strategy in Chapter 16, but what were the actual officers in charge on the night thinking?

Both national guidance and the London Fire Brigade's own policies envisioned a switch to an evacuation policy where stay put became untenable.[1] In fact, London Fire Brigade policy specifically stated: 'It may be necessary to undertake a partial or full evacuation in a residential building where a "stay put" policy is normally in place.' So why had the incident commanders not implemented this?

The most simple answer is that the plan to evacuate a building never went much further than the above line in the policy document. As such, incident commanders were not trained to evacuate buildings or to spot the signs that they might need to. As we've seen, Michael Dowden was left out of his depth and was never given the tools to oversee an evacuation. He understood that stay put was the policy for high-rise buildings and this is what he clung to. Andrew Walton, who briefly took over from Dowden, feared that residents would die if he ordered an evacuation of the tower. He believed that an evacuation was essentially impossible. He then handed over to Andrew O'Loughlin who was in charge until Andy Roe took over after his arrival at 2.30 a.m. He told the inquiry he had made an assumption that the fire was burning externally and not breaking back into the building. He was criticised in the inquiry report for not making more of an effort to answer this critical question – an answer easily available from those who were hearing calls from residents inside the tower.

But even if incident commanders had decided to evacuate the tower earlier, how would they have done it? Grenfell had no communal fire alarm to warn residents to leave. And how would residents with mobility issues would have escaped the tower given, as we will see, no plans had ever been made for their escape?

There is also the single staircase. This issue is sometimes overstated. One expert analysed the staircase and found that in normal conditions it was wide enough to accommodate the entire 293-person population of the tower, even if they entered simultaneously.[2] The time it would have taken for an able-bodied person to descend under normal conditions from the very top of the tower was four minutes. This shows us the extraordinary difference a decision to evacuate the building before the stairwell became smoke-logged might have made.

Nonetheless, one staircase did make evacuation harder. As soon as the door separating the lobby from Behailu Kebede's fourth floor flat was wedged open to run hoses, smoke was going to get into the stairwell. The firefighters had little option here, but it would have been a different story if the building had a second staircase. As the fire-fighting operation progressed, the stairs became busy with firefighters, equipment and water from hoses. Again, if a second staircase was present, it could have been left clear. British regulations do not insist on a second staircase and most tower blocks do not have one. Generally, this is a question of profit: the staircase takes away from the 'net lettable area', the amount of the building that can be sold. Britain is one of only two countries in the world (along with South Korea) not to require a second staircase in any buildings. In the United States and Ireland, one must be provided for all blocks four storeys or higher. In Canada it applies to any blocks of flats with two storeys or more. We are almost alone in the world in believing our tall buildings do not need a second stair. This has remained so even after Grenfell.

But even in the face of these difficulties, the number of people able to leave the tower unassisted, even as the conditions deteriorated, shows us that many of the lives lost may have been saved if they had been told clearly and unambiguously to get out of the tower when they or their relatives first called the fire brigade. But it had not happened. As the staircase and lobbies filled with smoke, an attempt at escape would mean a risk of dying in the dark.

*

On the twenty-first floor, Marcio Gomes phoned 999 for the second time at 2.46 a.m. Minutes before, he and his family had attempted to escape but had been forced to abandon their efforts by the smoke. He explained to the operator that the smoke was preventing the family from leaving the flat.

'Well, listen, the advice we're giving now you need to…'
'Yeah?'
'Leave, and you need to make your way…'

'I can't.'

'Down the stairwell.'

'We can't leave. There's too much smoke.'

'I know there's a lot of smoke outside, but you need to cover yourself in wet sheets, clothing.'

'We've done that. We can't – as soon as we open the door, the smoke just comes in. It's overpowering.'

The operator told him again that the advice was to leave – to put wet towels over his mouth and go. But with the smoke, the children and his wife's pregnancy it simply felt impossible.

'My wife is panicking. She can't do it… I've got young kids, and they're all scared, and, come on, this is ridiculous now.'

The operator assured Marcio that crews would be told about his position and sent to try and reach him, but 'if you can, if you can just give it one more try to get out, that is the best advice we're giving'.

It was a journey he did not think his pregnant wife or children could survive. He had been told help was coming and he decided the best thing to do was to wait. He remains frustrated that he was not told more clearly that it was unlikely firefighters would reach him. 'At no point did she [the call handler] say, no help was coming and we had no choice but to try and get out ourselves. If I knew that no help was coming, I would not have stayed in the burning tower with my family a minute longer,' he said in his witness statement.

Looking out of the window, he could see some people on the ground and hear shouting but the words were indistinct. In the flat directly above, another family were also trapped and they were hanging fairy lights out of the window to attract attention.

'All I wanted and was waiting for was the oxygen masks from the firefighters for my wife and [children],' he said. 'I followed the advice and information given to me. I believed that what I was doing was the best way of protecting my family.'

*

On the eleventh floor, Natasha Elcock had just gotten off the phone to the emergency services. The call had started at 2.44 a.m. and lasted for five minutes. It was around ten minutes after Joanne Smith had rescinded the stay put advice in the control centre, and during the call the then-incident commander Andy Roe made the same decision from outside the tower. But Natasha had not been told to get out. The call handler told the family to go into a room away from the smoke, block up any gaps and stay low. They moved to the bedroom nearest the door. The front door was visibly bulging with the heat from the fire.

She laid duvets out on the floor and told her daughter to lie down on them. 'By this point there was smoke in my bedroom but it was not so thick that I couldn't see. There was a gap between the wall and the window sill which I had complained about after the refurbishment as there was a draught and smoke was coming in through the gap,' she said.

*

On the top floor, the situation for the residents trapped was becoming increasingly desperate. In the flat where Sakina, Fatima and two others had taken shelter with a family of three, the fire was starting to break through the windows and the group decided to try and leave. Fire broke into the bedroom in the corridor by the front door. The man who lived in the flat shut the door, and the group sheltering there made the decision to leave. His wife handed out copies of the Qur'an and told them all to pray. He handed out wet towels.

It is not entirely clear what happened next, but in the dark and smoke the mother and grown-up son who lived in the flat were able to leave and make it down an agonising, brutal descent until they met firefighters and were helped out of the building. But no one else from the flat was able to leave.

For Sakina such an escape would have been incredibly difficult, given her physical condition. Her sister likely felt she could not leave her. Sakina also phoned her son Shahrokh at around this time. He heard his aunt say 'Forgive us', before the line was disconnected. They both died, along with the others who remained in the flat.

In the flat next door, Gloria Trevisan was on the phone to her mother. Earlier in the night, she and Marco had been calmer – expecting to be rescued by firefighters. Now, they knew they would not escape. By this point, her parents in Italy had also seen images of the fire on the television. They put the call on loudspeaker and listened – aware that it would likely be the last time they heard their daughter's voice.

The call to her parents lasted for twenty-two minutes. She told them that the windows were breaking, the glass shattering and the fire was coming in to the flat. She said the fire was too big and they thought the firefighters had stopped trying to come up. The smoke in the flat was making it hard for her to breathe or talk.

'She told me again what she felt for us and that we have to say goodbye and we had to be strong,' Gloria's mother recalled in a witness statement. 'At that point Gloria said she was cutting off the phone because she didn't want me to hear anything and she said goodbye to us. She wanted to cut off the phone because she didn't want me to hear her scream. She said she just wanted to faint and she wouldn't feel anything and said "I just want to stay with Marco now."'

The call cut off. Gloria's parents would never hear her voice again. She died with Marco and the mother and son who had come up the stairs to shelter in their flat at 1.30 a.m. The inquiry's smoke expert estimated they could have fled and survived, if encouraged to do so, until as late as 2.45 a.m. Of fifteen residents who ascended to the top floor, fourteen would die there.

'LET'S HOPE OUR LUCK HOLDS'

Cladding alone didn't cause the Grenfell Tower tragedy. The tower also had non-compliant lifts, a malfunctioning smoke control system and gas pipes which punched holes in the compartmentation at every floor. Its fire doors were drastically below the minimum standards. It had no plans for how to evacuate the many residents with disabilities who called it home. This all came down to how the building was managed. Many buildings still suffer the same management problems.

'THIS IS AN UNHAPPY CULTURE AND NEEDS TO CHANGE FOR THE BETTER'

The TMO's management of the tower dated back to 1996. This was a company owned by the council, which would take control of providing the day-to-day services to its 9,600 council homes. It was pitched as a way of giving residents more control over services.[1]

This organisation was not really a 'tenant management organisation', which are typically small, community-based organisations which spring up when residents on an estate strike an agreement to take direct control of some of the money paid in rent and arrange their own services. The TMO more closely resembled the arm's-length companies set up by many councils in the New Labour era to take over responsibility for managing their homes.

By 2009, thirteen years after it was established, the TMO was starting to seriously fail. Concerns from residents mounted to the extent that the council appointed an adjudicator to investigate: a Supreme Court solicitor, Maria Memoli. Ms Memoli published her report in 2008 and it was damning.

She reported a range of horror stories about the repairs service, including a tenant who had waited twenty years for a leak to be fixed. More broadly, Ms Memoli found issue with the culture at the TMO. 'Words such as "malevolent", "mistrust", "malaise" and "treated with contempt" were often used [by residents],' she wrote. 'This is an unhappy culture and needs to change for the better.'[2]

Following this report, a new chief executive, Robert Black, took over the TMO in 2009. He told the local press he would 'wipe the slate clean' at the organisation, which was described as 'racked by infighting and rebellion'. He promised a focus on improving repairs, to create cohesion on the board and change residents' perception of the organisation for the better. 'We need to rebuild that trust with people who are angry with us,' he said.[3]

Judith Blakeman, a Labour councillor for the area where Grenfell Tower is located and a former board member at the TMO, says there was some progress at first: 'Initially when Robert Black came in it did improve, but then it declined back to the pre-Memoli position,' she says. 'Across the borough, there was this huge neglect of everyday maintenance,' she recalls. 'The attitude to residents was "you are a nuisance, don't bother us with your repairs complaints".'[4]

This is borne out by the data. Across the entire borough, the general performance of the TMO in basic repairs was poor. It had outsourced the work of fixing issues with residents' homes to two private contractors, first Connaught, which went bust and then Morrison which replaced them. Both private companies struggled and in 2012 the TMO elected to set up its own repairs service: Repairs Direct, a company it would own. But this company too struggled with the basics. An enormous backlog of repairs jobs built up: 2,300 in 2015, 4,000 in 2016, 5,400 in 2017. Each of these represents a repair job to someone's home which was left undone. For an organisation which managed just 9,600 homes it is an astonishing record of failure.[5]

As the 2010s began, Grenfell Tower and the Lancaster West estate were running into disrepair. While many residents loved the tower and its community and took great pride in their own flats, the general state of the building was failing. Both lifts in the tower broke down regularly, leaving the tower's large cohort of disabled residents stranded inside. The communal boiler was reaching the end of its life. Hot water was shut off for whole weekends and sometimes drinking water too, with the TMO occasionally failing to even provide bottled water as a substitute. The tower was rocked by terrifying power surges which destroyed electrical appliances. Kitchens, bathrooms and windows were old and in dire need of replacement. Some residents had mould and mushrooms growing in kitchens due to persistent leaks and poor ventilation.[6]

For residents, trying to complain – or even report repairs – was a thankless task. In the early 2010s, an office on the estate where residents had previously reported problems was closed and replaced with a call centre. Residents describe being addressed in an 'abrupt and rude' manner when they reported complaints and felt the organisation was 'incapable of getting bookings finalised or contractors to arrive on the right day if at all'. Those who complained were forced to go through a three-stage process which rumbled on for months. Some gave up and lived with the disrepair. Others paid to have problems fixed privately or asked relatives and friends to do so.

Throughout this time, the TMO maintained a 95% satisfaction record for repairs. But evidence to the inquiry called this into doubt. 'The ethos was to select specific jobs which was almost a guarantee positive reaction to the works for example, a tap repair, to bolster the percentage,' a former employee told the inquiry. 'I tell you, if you walked round any TMO estate you would struggle to find anyone who had a good word to say about the TMO. They were experts at putting these stats out there and the reality for residents was completely different,' said Eddie Daffarn.

Amid the concerns that the building was not being properly maintained, fears mounted that the tower was no longer a safe place to live. These came to a head in 2010, when a fire started in one of the communal areas.

'LET'S HOPE OUR LUCK HOLDS AND THERE ARE NO FIRES IN THE MEANTIME'

Shah Ahmed and his wife Sayeda moved into a flat on the fifteenth floor of Grenfell Tower as council tenants in 1992. In 1999, they used the Right to Buy scheme to purchase their flat and become leaseholders. Shah had long been concerned about the standard of service he and the other residents were receiving and had lodged multiple complaints about windows, emergency lighting and the spiralling cost of the ancient communal heating system.

On the evening of 30 April 2010, Shah was outside Shepherd's Bush underground station, around a mile from the tower. He received a phone call warning him there was a fire in the tower and phoned Sayeda to make sure she was ok. When she opened the door to check the landing, thick smoke rolled into the flat. She screamed, 'Oh my God there is a fire and smoke everywhere, what shall I do?' Terrified, Shah told her to call 999 and hurried back to the tower 'crying like anything and running as fast I could'.[7]

The fire revealed a problem. The blaze had been on the sixth floor, but smoke had reached Sayeda on the fifteenth. Surely this was a danger? When Shah raised his worries with the TMO's representatives at a meeting he was told there was a 'minor fault' with the tower's smoke extraction system which had seen smoke leak out on upper floors.

But Shah was not satisfied. In September 2010, he wrote to the TMO outlining his fears. 'As you know fire does not kill as much as the effects of smoke and to our knowledge some of the residents nearly died due to smoke inhalation and suffocation,' he wrote. '[In] Grenfell Tower with its interior staircase and malfunctioning ventilation system there is certainly a high probability that in the event of another fire the whole building can become an inferno.' He was assured that the problems had been 'addressed'.

This was not true. In fact, the TMO had discovered that the vents on the tower's smoke control system did not properly close – meaning smoke could leak out. Instead of fixing this immediately, the TMO prevaricated for six years – trying to get the work completed as part of the wider refurbishment of the tower. This was despite the London Fire Brigade issuing a deficiency notice in March 2014, ordering them to fix the issue by May. The TMO refused a Freedom of Information request from Eddie Daffarn that would have exposed this – calling the details 'commercially sensitive'. In December 2014, after hearing the work was delayed again, the head of health and safety at the TMO, Janice Wray, wrote in an internal email: 'Let's hope our luck holds and there are no fires in the meantime.'[8] Replacement work would not start for another year. When it did, even the new system fitted did not comply with the building regulations.*

It was not the only fire safety warning that would be missed. Eddie's blogs, for example, repeatedly raised the

---

* The designers of the new system insist that it was impossible to provide a compliant system within the constraints of Grenfell Tower and their design was the best available option.

potential lack of access for fire engines down the small road which led to the tower, as the academy and leisure centre project cut off other access routes and replaced parking spaces. But the TMO did not like the blog. In fact, it was blocked from staff servers at their offices. Eddie's co-blogger Francis O'Connor was sent a legal letter by the council's lawyers calling it 'defamatory' and it was branded 'scaremongering' and 'agitating' in TMO emails. Notes from a the TMO board meeting in autumn 2015 also described Mr Daffarn and local Labour councillor Judith Blakeman, who was helping residents raise issues, as 'a negative force at Grenfell at present' and warned 'there is concern this unrest will spread to [neighbouring estate] Silchester' where a large demolition and rebuild was planned.[9] But Eddie was right. On the night of the blaze, seventy fire engines would be backed up down this street and around the estate.

As we noted in Chapter 10, the refurbishment project resulted in many more complaints from residents and calls for investigation of the safety of the work. But the TMO shut down public meetings, initially refused to meet with the group Eddie set up and did not involve residents in decision making. They were never told, for example, of the decision to switch the cladding for a cheaper option. It is not unreasonable to think that Eddie or Shah might have taken the step of Googling 'ACM' and discovering the reasonably well-publicised fire risks, and the blazes that occurred in the Middle East. But instead of being invited to participate in the planning, they were shut down and ignored. And amid all the problems with repairs and maintenance, there was one particularly critical failure that was never addressed: the fire doors.

DEFECTIVE DOORS

The whole principle of high-rise safety in this country is wagered on the idea of 'compartmentation'. This means that when a fire breaks out in one flat, the smoke and flames should stay contained within that single box, rather than spreading throughout the building. It is this that justifies the stay put policy. We have seen in the prior chapters how this was undermined from the outside by the addition of combustible plastic to the walls of thousands of homes around the country. But it was also being undermined from within the building.

Compartmentation relies on well-built and well-maintained buildings. Gaps which allow smoke to spread from one 'compartment' to another mean it will fail. Unfortunately, the maintenance and construction of buildings is too frequently not done with fire safety in mind. This means the necessary protections are sometimes missing from the start, or removed as the building is altered. The architect Sam Webb, for example, has told me about one building he inspected where the wires from the door entry phone 'had been converted into a 16-storey chimney by someone with a Kango Hammer just punching a series of holes through concrete slabs.'[10] Walk around any older council high-rise and you will likely notice gaps and cracks, sometimes clumsily filled up with plastic foam. As Brian Robinson, president of the Fire Sector Federation told *Inside Housing* in 2018: 'I think the issue with compartmentation is a real standard one. If you get plumbers, electricians, you name it, coming into a building the first thought on their mind isn't fire safety. It's actually, "Let's get the job done", and the worst thing that can happen is they punch a hole in the wall and then fill it up with magic foam – which is flammable. It's legion.'[11]

Flat entrance doors are an utterly crucial feature in this system. Fire is most likely to start inside a flat, and the door is all that is stopping the smoke from pouring out onto the landing – preventing anyone else from escaping and hindering the firefighters sent in to tackle the blaze. This means the door must be resistant to both fire and smoke leakage.

The doors in Grenfell Tower were relatively new. The TMO had carried out a major programme of door replacement across its stock from 2011 onwards and had fitted new doors made by Manse Masterdor. These doors were plastic: a glossy shell with foam insulation in the centre.

This is a modern innovation – historically most fire doors are solid timber slabs which will resist fire for a long period of time. But 'composite' doors are less reliable. Those sold for use at Grenfell were supposed to meet the minimum legal standard of thirty minutes' fire resistance. They were advertised and sold as doing so. But the doors in fact had major problems. They had glass panels and different locks, hinges and letter plates. These may sound insignificant but can substantially reduce the fire performance of the door. Investigations after the fire showed a substantial number had been destroyed by the fire and when the Metropolitan Police tested undamaged doors they failed in fifteen minutes – half the time required for legal compliance.[12]

The inadequacy of the fire doors was only one problem. Even a defective fire door will provide a degree of protection if it is closed. At Grenfell, many were open. Residents had repeatedly warned the TMO and council about this.

A SELF-CLOSED CASE

Closing the door behind you is not a priority when fleeing a fire. This is why, legally, fire doors must be fitted with self-closers to ensure they swing shut when the resident leaves.

The doors at Grenfell had self-closers, but they were defective. It was not long after the doors were installed that residents began reporting problems. Sometimes the device – a small chain contained inside the door hinge – would hold the door open making it impossible to close. In other instances they would snap it shut too quickly. These issues were reported to the TMO. But instead of fixing them, the broken self-closers were simply removed.

'My front door was replaced around 2013. I remember that not long after it was installed – a few weeks I think – the door got jammed and would not shut. I called the TMO to report that my door was broken,' a nurse who lived on the seventh floor told the inquiry. 'It took around an hour for someone to arrive to fix it. The person that came adjusted the mechanism on the back of the door which had previously made it close automatically, so that it no longer worked and my door remained open, not closing automatically. I remember asking whether this was a problem. The person that did the work told me it was not.'[13]

Natasha Elcock said her fire door broke shortly after it was installed. 'I was going out and it literally just dropped off its hinges,' she said in evidence to the inquiry. She said that she reported the fault to the TMO, which sent a workman around to fix the issue. 'Someone came out and took out the self-closing mechanism and they never came back,' Ms Elcock said. 'Once he took it out, the door no longer closed by itself... He said someone would be back to replace it, but that never happened.'[14]

There were many similar accounts. In August 2015, Eddie Daffarn tried to raise this problem of self-closers with the TMO. The flat next to him had been left vacant when the resident moved out and he was frustrated that Rydon staff had left the door standing open after they entered to carry out works. When he tried to pull the door closed, he realised the device was preventing it closing. He reported this, but later received an email from Peter Maddison asserting that the door could have been closed if he had pulled it harder. He angrily rejected this as 'a complete pack of lies'.[15]

Had the TMO investigated more closely, it would have discovered an endemic issue. Seamus Dunlea, a handyman who worked at Grenfell Tower and died before the start of the inquiry, explained what had been happening in a witness statement. He said that there was a simple problem with the closing device. 'Some of the screws were pulling out of the door and some were pulling out of the frame,' he wrote. 'You can't reattach it because of the fixings so I would have no alternative but to remove it.'[16] He claims to have shown one of the closers he removed to a manager in the TMO offices and explained the issue, but says nothing was done. 'I got no response from management... so all I could do was remove the door closer. That made an illegal door because with the closer pulled out, the door wouldn't self-shut,' he said.

Emails show that senior managers at the TMO were told about this in December 2015 and told him to stop removing the devices.[17] But there is no evidence of any work to check the doors he had disconnected or fix them. And there were similar problems elsewhere in Kensington.

On 31 October 2015, there was a fire at Adair Tower, a block around a mile from Grenfell Tower, also managed by the TMO. This was a serious fire, which resulted in smoke inhalation injuries to residents and a full evacuation of the building by the London Fire Brigade. Sixteen residents were treated at hospital and twelve flats were left 'uninhabitable'.[18]

At the time of this fire, Adair Tower did not have self-closers fitted on its flat-entrance doors. In fact, on 12 October, nineteen days before the fire, the London Fire Brigade had issued the TMO with a 'deficiency notice' demanding that all doors be fitted with self-closers.

But some senior figures at the TMO and RBKC were reluctant to prioritise this work – even after the fire. Mr Maddison wrote in an email in November 2015 that retrofitting self-closers was 'not a statutory requirement' and suggested the work be downgraded from high priority to 'low or advice' in the TMO's internal risk assessment.[19]

In December 2015, the London Fire Brigade stepped up the pressure with an enforcement notice demanding the work be done. But the council did not want the bill. Notes of a meeting from February 2016 say: 'RBKC do not want to do this work if not required.' Internal emails show Laura Johnson, director of housing at RBKC, complaining to Robert Black that the LFB had forced them to repair the fire doors at Adair, calling the fire brigade 'unaccountable' and 'hard if not impossible to challenge'.[20]

But the LFB was not just worried about Adair. It told the TMO that it expected all fire doors within the 650 tower blocks managed by the TMO to have self-closers, and threatened more deficiency and enforcement notices if they did not. It also wanted to see the introduction of an annual monitoring programme to keep track of their condition. The cost of retrofitting self-closers was estimated at £620,000 and there would be an ongoing cost of around £200,000 a year associated with monitoring. For a borough with the rents paid by 9,000 homes available to spend on their upkeep, this was a very minor sum.

But RBKC did not want to spend this money. Minutes from October 2015 say: 'LFB are putting pressure on us to fit door closers on all our stock across the business. Laura Johnson has said no to this.'* Instead, to 'make funding the programme more manageable', the council gave itself five years to fit self-closers instead of the one-year turnaround the LFB had hoped for.

It also did not carry out the inspections the LFB had asked for. TMO minutes from March 2017 note that an inspection regime would be 'a huge cost' and said 'nobody [else in the social housing sector] has an inspection process at present'. As a result, it was agreed to 'hold off' introducing one.[21]

The LFB, though, was not satisfied. In November 2016, it carried out an inspection at Grenfell Tower and issued a deficiency notice. This identified two doors – those to Flat 44 and Flat 153 which did not self-close. 'The protected route [the staircase and the lobbies] has been compromised by the fitting of doors that do not self-close,' it said. It gave the TMO until May 2017 to fix them.

This work was not done. Investigations after the fire showed 43 of the 129 doors in the tower had no self-closers installed and 34 that did were not working properly. As residents fled their flats, smoke poured into lobbies through the open doors. This was described as a 'key event' in the cause of the fatalities by an inquiry expert, because

---

* Inquiry transcript, 12 May 2021. Ms Johnson denied saying no, but claimed she simply wanted a report clarifying exactly what was required.

it discouraged residents who opened their front doors from fleeing. The tower's exterior was made flammable by the cladding. Its interior was rendered unsafe by neglect, penny pinching and a refusal to listen to complaints.

THIS 'WILL LEAVE TENANTS...WITHOUT THE PROTECTION THEY NEED AGAINST INCOMPETENT LANDLORDS'

The failures of fire doors go much wider than Grenfell Tower. A fire door inspection service warned in 2022 of a 'tragedy waiting to happen' after revealing that 75% of more than 100,000 doors it inspected were defective. Even five years on from the fire very few social landlords have active, regular inspections in place.[22]

After the fire, testing discovered 'a performance issue with [plastic] composite 30 minute fire doors across the market'. Sales of composite doors had to be suspended, after twenty-five other model samples from a range of suppliers also failed to resist fire, some in less than ten minutes.[23] There has been no government-led effort to ensure they are replaced, with the decision about what to do left to the discretion of landlords.[24] Some of this widespread failure comes down to a familiar story about deregulation. 'We've spent years campaigning that there is a systemic issue,' Iain McIlwee, chief executive of the British Woodworking Federation, said in 2018. '[But] we don't tend to have pre-scriptive regulation in the UK. They [the government] may have seen it as a burden on business.'[25]

After the Lakanal House fire in 2009, the issue of fire doors was specifically raised with Brian Martin, the civil servant who also missed critical warnings relating to cladding. At an industry liaison group to discuss issues arising from the fire a few months after the fire, participants drew attention to problems which had emerged at a block of refurbished flats in Hammersmith and Fulham. Serious problems with fire doors had been uncovered. When tested, the fire doors throughout the building failed – some in just seven minutes. The meeting noted that test laboratories did not disclose data to third parties and main contractors 'do not care and are not interested'. 'It was agreed that the whole affair was highly disturbing,' the minutes note. But Mr Martin simply told the meeting that while some of these issues were 'under consideration' there would be no major review of the guidance until 2016. As we have seen, this was never done.[26]

Since the Grenfell Tower fire, the issue of how complaints of social housing tenants are too often dismissed has gained urgency, thanks to pioneering investigative journalism from ITV News and campaigners such as Kwajo Tweneboa, highlighting horrendous conditions.[27]

Some of this comes down to financial decisions. We built social housing in vast numbers after World War II and then neglected to spend sufficient money maintaining it. But poor conditions are one thing. Another issue is that residents have little recourse if their landlord ignores their complaints. Like those at Grenfell Tower, they are forced through the dizzying, exhausting, time-consuming process of internal appeals which sometimes appears designed to exhaust and frustrate complaints.

Before Grenfell, there was no regulator overseeing the housing management of the council and housing association landlords responsible for millions of people's homes and safety. The Labour government had begun the process of setting one up under Gordon Brown. But the body it created – the Tenant Services Authority – was axed by the coalition government when it took power in 2010, before it had even seriously got going as part of the 'bonfire' which saw a vast array of state institutions unceremoniously abolished. 'The TSA is toast,' boasted Grant Shapps, then housing minister, to a reporter before announcing its axing.[28] Professor Martin Cave, the academic who had advised the creation of the Tenant Services Authority, warned the move 'will leave tenants and groups of tenants without the protection they need against incompetent landlords.'[29]

Instead, the government set up a regulator which would only really consider the financial performance and governance arrangements of social landlords. This was due to the need to attract private finance into the social housing sector to compensate for the cuts being made to government grant.

The lack of effective regulation was an issue which was becoming increasingly prominent in the months before Grenfell, due in part to a repairs scandal which blew up over the actions of a large housing association called Circle. Residents suffered horrendous conditions but the regulator told them in summer 2016 that it would not be 'proportionate' to get involved. The local MP raised their complaint in Parliament. 'I accept the logic behind [the government's] position on deregulation, but what is the cost in terms of oversight and accountability?' he asked then housing minister Gavin Barwell in January 2016.[30] The country would soon see.

'I DEARLY MISS OUR COMMUNITY'

There are a couple more points worth noting about the management of Grenfell Tower. The first relates to race. To what extent did the ethnic make-up of the tower affect the treatment of residents? On one level, this point is practical. With the complaints system complex and difficult to navigate, and collective groups shut down, those who did not have English as a first language were disadvantaged – an issue which impacts social housing blocks across many inner city areas.

Grenfell Tower was overwhelmingly occupied by people of colour. On the night of the fire, 85% of those who died were not white. In a written statement, lawyers for one group of bereaved and survivors asked 'whether there is any link between the failure to maintain Grenfell to an adequate standard and the fact that Black and minority ethnic people were disproportionately concentrated in Grenfell Tower'.[31] Emma Dent Coad, the former Labour MP from the area, has described hearing council workers refer to the Grenfell Tower area as 'little Africa' and say it was 'full of people from the Tropics'.[32]

Certainly, the racial divide in Kensington is stark. A white Brit from Knightsbridge can expect to live to ninety-one while a Moroccan man living three miles away can expect to work his whole life and die at sixty-four, never having been able to claim his pension.[33] As stated, the council is primarily elected by the richer, white residents of the borough. Would it have allowed the kind of conditions endured by the residents of Grenfell Tower to exist in a tower with white residents? This is a question many bereaved and survivors want the inquiry to answer.

Despite the management issues, many residents of Grenfell Tower loved their home. This is something that has come through clearly to me in all the interviews I have done since the fire. Yes, the lifts broke down, it could be cold and the communal heating system was poor. But this was a place with a thriving, happy community who wanted to live there. The story of Grenfell Tower is not an indictment of social housing, or even tower blocks. It is a call to take better care of these precious assets that form the backbone of so many communities. This is sometimes lost in reactions to the tragedy, which portray Grenfell as a miserable place to live, even before the fire took hold. This could not be further from the truth of those I have spoken to, and many who have spoken to the inquiry. Eddie Daffarn submitted an eloquently written 127-page witness statement to the inquiry. He concluded it with these words: 'I dearly miss our community. We came together in the face of adversity before during and after the fire. We were not just neighbours. Since the fire there are people out there who have said terrible things about our community, things that are so far away from the reality of what it was actually like, that it has really hurt. We will never have the chance to really show people what that community was like. That thought is truly heart-breaking.'[34]

13

3 A.M.

On the walls of Grenfell Tower, the fire had now crept around the corner to the western face of the building – finding yet more deadly plastic to feed its march around the building. The clockwise and anti-clockwise spread of the flame were closing in on each other – a pincer movement for the few flats not yet on fire.

Flames now covered two sides almost entirely and spread downwards on the others in a jagged line of white heat, with the orange glow of fierce blazes dotting many windows up and down the tower. At this stage, just over a hundred residents remained in the tower – some alive and some already dead. The heat was rising in the stairwell, and many of the firefighters had been working for hours.

'A lot of the firefighters returning from their efforts were extremely exhausted, passing out, we were trying to get them water, take off their tunics, get them oxygen. They were passing out, certainly suffering from heat stress,' Louisa De Silvo, a senior firefighter who helped run the operation from the bridgehead, told the inquiry. This made getting accurate information about where they had been and what they had seen almost impossible. Firefighters were also being asked to head up the tower without any water to fight the now rampant fires in the flats they were entering. 'Any firefighter going near or into a fire will take water. That's our baseline for our own safety,' Ms De Silvo added. 'I remember having to say to him [a firefighter] that it was unlikely there would be water and that he was to try to effect rescues… That's significant to me because it was putting them at so much risk.'

One firefighter deployed to the tenth floor at around this time described the heat as like 'a blast furnace'. 'Every time you tried to get up to kneel you were getting your head cooked,' he says. 'You could feel your head pounding in your helmet, with your heartbeat and the adrenaline, but you could actually feel your head banging with the heat.'

One crew of four had been sent up to the twelfth floor to find a missing firefighter. While they were searching the floor, the firefighter was found so the mission was changed – two remained on the twelfth floor to carry out a search and rescue for any residents trapped. As they searched, two women came out from one of the flats and shouted for their help. With the help of the two firefighters, the women were led to safety and survived. The fire was spreading inside their flat by this point and there was little time left – they were likely only minutes from death when the firefighters arrived.

*

Just after 3 a.m., Natasha Elcock made her first attempt to escape from her home on the eleventh floor. But when her partner opened the front door, a huge wave of smoke flooded in to the flat and set off the smoke alarm. Natasha could see the lobby was still black and the heat was intense.

She phoned 999 again and again – pleading with the call handler to send someone to rescue her family. One operator told her that the brigade were 'getting the fire under control'. 'She said that if I thought I could stay in my flat until they come and get me, I should but if it gets so bad then we needed to run for it,' Natasha recalled in a witness statement. 'I begged her to tell them to do every-thing they could. She told me I needed to stay calm for my daughter.' This was after the stay put advice had been lifted. Natasha should have been told unambiguously to get out.

It was getting hotter and smokier in the flat. Her partner had already extinguished a blaze in one of the bedrooms. The window panel outside her flat was on fire. Natasha no longer expected firefighters to reach her. She thought she would die. But she resolved not to give up, for the sake of her children.

*

The night before the fire, Rania Ibrahim had spoken to her sister Rasha in Egypt. The women laughed together as they always did, but Rasha found Rania's tone strange. She told her not to worry, to take care of her health and not to be upset over anything. 'I had a strange feeling as if she was saying her goodbyes and as if she was departing,' recalls Rasha.

Early in the morning on 14 June, Rasha rose to prepare her early morning breakfast, ahead of a day of fasting as it was Ramadan. She got a phone call from another of her sisters telling her Rania's flat in London was on fire. 'That was the most difficult phone call I have ever received in my life,' she said. 'I knew she was alone with her daughters and she gets scared quickly. I tried calling her but couldn't get through. I watched live TV as the fire was engulfing the whole tower as if it was a carton box and I smashed the TV screen as I wished I could reach out and get my sister out of the building.'

Earlier, Rania had posted footage from her flat onto Facebook Live, but there had been no more posts. Rasha began posting frantically in English and Arabic: 'Where is Rania'. She was contacted by people from all over the world offering their sympathy and encouragement. 'I cried and they cried with me over the phone. They assured me and gave me hope that Rania is well and that this pour of love from all over the world meant she was well,' Rasha recalls.[1]

But Rania was not well. Shortly after 3 a.m., she phoned a close friend who was outside the tower. She was coughing and struggling to breathe. Her friend implored her to leave the building, but Rania said she had been told to stay where she was and wait for help. She believed the police helicopter circling the building might be used to rescue them from the roof. Rania was inside Flat 203 with several residents who had fled from the lower floors and also her daughters, Fethia and Hania. Fethia, aged four, was described as a confident child with a strong and vibrant personality was a role model to her younger sister, aged three, who was described as quieter, but wise beyond her years. They had inherited Rania's spirit. Rasha described her sister as 'a beautiful soul'. 'No one could sit with Rania and not smile,' she said. Hundreds of miles away in Egypt, this is all Rasha wanted. To sit with her sister again and smile. But she would not. Rania spoke to a friend and her sister outside the tower at around 3 a.m. The children could be heard in the background coughing and asking for their father. Rania still believed firefighters were coming and that she should stay put and wait for them. But no one was coming. The highest any firefighter reached was the exit door to the twentieth floor, where they found a trapped resident and took her down. No crews ever reached the top floor. After 3 a.m., no one would hear from Rania, or the others sheltering in Flat 203 again.

*

At 3.20 a.m., incident commander Andy Roe held a tactical meeting inside one of the command units near the burning tower – bringing together senior figures from the three emergency services and a representative from the Royal Borough of Kensington and Chelsea. He asked the police to establish a cordon to keep the crowd back from the tower, and he also wanted plans of the building from the council officer – which they were unable to provide. By this point, he was so worried about the structure of the building that he had made an urgent call for a structural engineer to attend and advise whether or not it was at risk of collapse. The shadow of 9/11 was undoubtedly in the minds of the firefighters present.

Following this meeting, he decided he needed to enter the tower himself to assess the conditions – essentially with a view to making a call about whether or not it was still 'tenable' to carry on deploying firefighters to rescue residents. He briefly checked on the status of calls from trapped residents in the other command unit. By now a more orderly 'grid' system laying out all the flats and the details of those trapped had been drawn up on one of the white boards. He was told an estimated 100 people were still stuck inside the tower. He knew he would face a devastating choice: keep sending firefighters up the tower, potentially to their deaths if the tower collapsed, or abandon trapped residents waiting for help. He left the command unit and approached the building for the first time.

As Roe was about to enter the tower, the body of a man came crashing down in front of him, striking a firefighter. It was a resident who had fallen from an upper floor. He instructed crews to remove the body and ensured the firefighter who had been struck was sent to the ambulance services. Then, ducking under a riot shield, he entered the tower.

Inside, he observed the system: details of trapped residents scribbled on the walls and firefighters waiting to be deployed up the tower. He made a decision. He ordered every piece of extended duration breathing apparatus in London to be delivered to the tower. The rescue operation would continue. Before leaving, he paused to give a speech to his officers.

The speech, in the context of the fire service, was extraordinary. He told them that they would have to do what they could to save lives – and this would mean putting brigade policy to one side and taking risks they would not normally consider. The situation demanded more. People were trapped and there was a 'moral obligation' to save them.

In their witness statements, firefighters present described it as a 'rousing' speech.

'The essence of what he said was, "look lads, this is a once in a lifetime incident, we've never faced anything like this before, and it goes outside brigade policy. Do your best." I think it really helped and those younger in service really liked it. Everyone was incredibly motivated,' recalls one. 'Ordinarily we stick religiously to policy but because of the situation, this went totally outside of that in order to get as many people out of the building as we could.'

*

On the twenty-first floor, Marcio Gomes made his final 999 call at 3.25 a.m. The fire had now reached their flat. It was no longer possible to stand in the flat due to the smoke. Marcio saw flames outside the bedroom window. They spread into the room and began to burn the curtains and spread along the ceiling as he was on the phone.

'Oh shit, the fire's here, the fire's here,' he said to the operator.

'The fire's in the house now. Right, right so if the fire is in the house you need to get out of the house, don't stay in the house,' the operator replied.

He pulled the bedroom door shut and shouted to his wife and Helen that it was time to leave. They had wet tea towels already and Marcio told all the girls to hold hands.

'Right, you've just got to go,' said the operator. 'Stay with me on the phone, okay.'

He had told everyone to go in a straight line to the stairwell, grab the handrail and not let go until they were out of the tower. 'Hold hands, go straight. Hold hands go straight to the stairs. Get to the stairs now! Go!'

'Just all stay together and keep the children, make sure you keep hold of the children as well,' she said to Marcio. 'Don't talk to me, just talk to them.'

It was so dark on the stairs that Marcio could not see a thing. The smoke was awful: every time he tried to breathe he felt like gagging and throwing up. 'The smell of the smoke was unnatural and toxic, it smelt of burning chemicals. It was very hot, but I had a cold wet towel over my head… The tea towel helped initially but the soot, smog and tar was getting through the tea towel making it difficult to breathe through it,' he recalled in his witness statement.

'They're going along the corridor. He's got the three kids with him. His wife's pregnant,' the call handler says.

Marcio described the heat in the stairwell as 'like the heat that comes out of an oven and blasts you in the face'. Over and over again, Marcio shouted the words 'keep going girls' to the children in front of him. 'I felt that if I stopped encouraging them and pushing them to continue, then they would simply halt and that we would all die on that stairwell.'

'Just keep going, mate, you're a top bloke,' said the call handler. 'Keep going, this is really important, just keep going, all right, you're a top man.'

As they kept on, picking their way down, desperate for air, he heard the voice of one of his daughters behind him. 'Daddy, I can't go on anymore, I can't.'

He didn't know how she had got behind him. 'Follow my voice,' he said, trying to find her in the dark. He kept trying to encourage them. 'Come on girls, keep going, keep going. Keep a hold of rail,' he says. 'Girls! Come on, girls, keep holding the rail, come on, push, push through.

'Girls! Come on, girls, please. Please.'

'I tell you, anyone would like a dad like you, anybody,' said the call handler.

'Girls? Oh, fuck, I can't even hear them now.'

He carried on calling out for them, sobbing.

'Oh fuck, I need my wife. Girls! Oh girls, please (crying). Oh my girls, oh fuck.'

'Right, the firemen are aware of where you are, okay?'

'I can't leave them. I need to go upstairs, I need to go up.'

'You need to go and find —'

'I need to find them.'

He told the operator he was going up the stairs and she advised him to stay low, gather his breath and do it a bit at a time.

'Oh please, please God, please God (inaudible), don't (inaudible) my girls. Take me, not my girls (crying). Please, please take me. Why, why me? Why did you take them?'

Suddenly he saw a light and raced down towards it. He found a firefighter and began pleading with him to help find his wife and the girls. He pleaded with the firefighter to let him go back up. 'I need to get my daughter, my daughter's up there. Come on. I need to go and find them. I need my daughters and my wife and she's seven months pregnant. Oh come on.'

Then the sound cuts out. Marcio had reached the fourth floor. He waited.

And then the firefighters came past carrying his daughters. Andreia made it out as well. The family were reunited outside the tower.

14

RISK ASSESSMENT

The process of fire risk assessing residential buildings in England is woefully flawed: often little more than a form-filling exercise by an unqualified and unregulated consultant. This is a problem with deep roots. The Fire Precautions Act 1971, which followed a fire in a hotel and pub in Essex which killed eleven people, required certain buildings to receive a licence following rigorous inspection by the fire service.

This Act covered hotels and workplaces, but never applied to residential blocks. The legislation could have been extended to high-rise flats, but doing so was left to the secretary of state's discretion and it was never done, amid fear that the costs of keeping the properties fire safe would limit the supply of social housing. 'There may well be cases... where the risk of fire is more acceptable than the risk of homelessness,'[1] wrote a minister in 1970.

This system stayed in place until 2005 when it was overhauled by Tony Blair's government. The Regulatory Reform (Fire Safety) Order changed the emphasis. Rather than receive a licence from the fire service, building owners were required to assess the risks themselves. They were required to arrange a 'suitable and sufficient' risk assessment of their premises and fix any problems identified. Blocks of flats were included in this new regime, an accidental consequence of EU law which required risk assessment for all workplaces.[2]

These assessments were limited. Most building owners took them to apply only to the communal areas where their workers would enter – the stairwells, corridors, lobbies. Crucially, there was no restriction on who should carry out the inspections, other than that they should be done 'regularly' with the support of a 'competent person'. It was up to the building owner to decide what this meant. The result, unfortunately, has been to push them towards the cheapest option. There have been regular calls to overhaul this system and at least have a mandatory legal level of qualification and third party approval. But government after government has rejected this.

'Businesses are unlikely to welcome the burden of further regulation in this area,' wrote civil servant Louise Upton in July 2010, in one of many examples of this.[3]

Throughout the 2010s, there were repeated warnings that the competency of fire risk assessors were not up to scratch – including from the London Fire Brigade. But civil servants continued to assert that imposing standards would result in 'unnecessarily high cost burden on business'.

There were other problems: placing building owners in charge of the process created a perverse disincentive not to notice problems they would then have to pay to fix. In 2022, I spoke to a risk assessor for a variety of social landlords who described being 'bullied' for raising concerns and being 'placed under huge amounts of pressure to change reports'.[4] Another piece of research showed virtually all of the assessments in social housing blocks were 'type one': the most basic form which involves no intrusive checks and can overlook major problems hidden beneath surfaces.[5]

It was this flawed, broken system that was charged with spotting the problems with Grenfell Tower before it burst into flames. It is no real surprise that it failed. But the manner of the failure is still disturbing.

HE IS 'WILLING TO CHALLENGE THE FIRE BRIGADE ON OUR BEHALF IF HE CONSIDERED THEIR REQUIREMENTS TO BE EXCESSIVE'

In 2009, four years after the legislation requiring building owners to carry out risk assessments had come into force, the TMO was simply relying on in-house staff to carry out this work. They did not have formal qualifications and while this was potentially acceptable under the terms of the legislation, the London Fire Brigade did not like it. They told the organisation to appoint a competent risk assessor or face an enforcement notice.

As a result, the TMO procured Salvus – a reasonably well-known fire consultancy – to assess its higher risk properties within six months. Salvus also produced a management report which raised serious concerns about the TMO's general management of fire safety.[6]

In February 2010, Janice Wray, head of health and safety at the TMO, decided to dispense with Salvus' services and procure a new risk assessor. In an email to colleagues she described the consultancy as 'very rule bound'. 'Despite what they say about being prepared to challenge the LFB and acting on our behalf as we are their client, I believe they have shown some reluctance to challenge the LFB on thorny issues,' she wrote.*

One of the risk assessors engaged by Salvus at this point was Carl Stokes, who had retired from nineteen years in the fire service in September 2009 and set himself up as a risk assessor. He had carried out a couple of short training courses and had audited other people's risk assessments in his role as a firefighter but had no prior experience of doing his own. Despite this, his CV claimed he had 'experience of undertaking risk assessments' and suggested an impressive list of qualifications with the inclusion of nominals describing himself as Mr C Stokes ACIArb, FPA Dip FP (Europe), Fire Eng (FPA) and so on. But many of these were in fact a mix of the irrelevant (an arbitration qualification) and the non-existent, e.g. Fire Eng (FPA). Mr Stokes denied trying to deceive his clients when asked about these at the inquiry, saying he was simply trying to list training courses he had attended.**

---

* Wray, evidence 7 June 2021. Ms Wray claimed she was frustrated that Salvus had 'sat on the fence' regarding a specific block where the London Fire Brigade wanted a dry-riser retrofitted.

** Inquiry transcript, 25 May 2021. An expert witness, Colin Todd, said Mr Stokes' qualifications were 'suitable for the purpose of carrying out [risk assessments] for Grenfell Tower'.

Mr Stokes set himself up as a sole trader and applied to take over responsibility for the TMOs 'medium risk' blocks. He was appointed in September 2010 and would go on to take over all of its fire risk assessment work for the next six and a half years. TMO board papers in 2010 record that he had offered 'the most competitive price' for the work and had also proved himself 'willing to challenge the fire brigade on our behalf if he considered their requirements to be excessive'.[7]

The inquiry saw that he occasionally copied and pasted from one report to another, when reporting on different buildings. This resulted in some embarrassing mistakes. For example, his risk assessment of Grenfell Tower in April 2016 said a visual inspection of pigeon netting around balconies showed it was well-fitted and not damaged. Grenfell had neither pigeon netting nor balconies.[8]

When his assessments discovered issues, he produced a list of 'significant findings' that were compiled into a list for the TMO to address, colour-coded to indicate their priority. But the TMO was not particularly efficient at getting them done.

By March 2014, board papers record an astonishing backlog of 1,400 incomplete actions in the 650 blocks that required risk assessments, each representing a known and unaddressed risk to the safety of residents.* Board papers

---

* Inquiry transcript, 10 May 2021. KCTMO witnesses at the inquiry linked this to the poor performance of the private contractors to whom they had outsourced repairs in 2012, as well as the complexity of completing some larger actions.

suggest the TMO decided not to disclose this backlog to the London Fire Brigade as it 'would result in more scrutiny from the LFB and also possible enforcement action'.[9]

Over the next three years, the TMO would try to reduce this backlog, regularly discussing it in committee meetings. But it would never get it completely under control. The day before the fire, there were still 287 actions outstanding, 128 of them more than a year old. There were unaddressed fire risks throughout the homes managed by the TMO. And Grenfell Tower was no exception.

Carl Stokes assessed Grenfell Tower five times. He was supposed to check flat entrance doors, but was not contracted to look inside the flats, so his inspections were primarily visual and carried out from the corridor. Occasionally, if he saw a resident he would ask them to show him their door. He claims to have reviewed around 5% of the tower's doors as part of these ad hoc checks.[10] But he missed the fact that so many of them did not self-close.

He also considered the issue of whether people with disabilities were safe in the tower. In his five risk assessments for Grenfell Tower, he invariably ticked a box saying the building was 'provided with reasonable arrangements for means of escape of disabled people'. In commentary next to it, he wrote that there was 'no evidence of any resident within the premises who suffers from sensory impairment that would prevent them hearing a shouted warning of fire'. These statements remained in his assessments, despite clear evidence of disabled residents, including emails referring to a blind resident, and the presence of a mobility scooter being raised in one of his own assessments. The same statement was copied into assessments of other blocks he reviewed for the TMO.*

In December 2012, the TMO was contacted by the London Fire Brigade about a pilot it was planning to run which would fit small sprinklers into the homes of disabled residents to minimise the risk of fire. The email was forwarded to Mr Stokes. He replied: 'I would say you have nobody that this refers to because if you had you would have undertaken [evacuation plans] for them and implimented [sic] any findings which would have included additional fire safety measures. If you identify anybody now questions like why were they not including in the buildings' FRA [Fire Risk Assessment] spring to mind. A good response I believe would be thank you for this information if we find anyone in the future we will let you know.'

Asked about this at the inquiry, Mr Stokes denied he was effectively telling the TMO to 'keep quiet' about the presence of disabled people in their tower blocks and the lack of any plans for their evacuation.**

---

* Inquiry transcript, 26 May 2021. Mr Stokes said that since KCTMO did not alert him to the presence of any residents with disabilities, he was content to assume they were not any present in the building. An expert witness Mr Todd said he would not have expected the assessor to have any details about disabled residents, beyond high level information provided by the building owner.

** Ibid. Mr Stokes said he believed the LFB were specifically interested in heavy smokers, and he was not aware of any of these cases in Grenfell Tower.

What about the deadly cladding on the walls of the tower? Mr Stokes was asked by the TMO to carry out two reviews of the refurbishment works. In April 2016, he recorded in his risk assessment that the panels were 'fire rated' and said the system had been 'approved and accepted' by building control officers. Mr Stokes said that he'd had 'informal conversations' with figures from Rydon when they showed him around the site and gave him 'assurances' that building control were happy with the cladding. Handwritten notes following this discussion described the cladding as 'Ok FR no timber' and 'cladding external non-combustible'.*

In April 2017, following the fire in Shepherd's Bush the year before which had prompted Eddie Daffarn's warning on his blog, the LFB wrote to all London councils with a warning about cladding, advising them to check it as part of their fire risk assessments. When the letter was received at RBKC, it was forwarded to the TMO's head of health safety Janice Wray, who forwarded it in turn to Mr Stokes. 'My understanding is that we do not have any blocks with external cladding of this nature. Are you able to confirm please?' she wrote on 24 April.

He replied: 'Grenfell was clad but the cladding complied with the requirements of the Building Regulations, lots of questions asked of Rydons [sic] and answers received back from them.' He would later accept that he made no

---

* Inquiry transcript, 1 June 2021. Expert witness Mr Todd said that Mr Stokes had 'absolutely gone far enough' in making checks, if he had assured himself that they were approved by building control.

checks of the records before offering this assurance, was not a construction materials specialist and was not qualified to give such an assurance.[11] Three months after this exchange, the truth would be unequivocal.

15

4 A.M.

Dawn was starting to break in west London. At 4.08 a.m., the two lines of flame that had been spreading in opposite directions around the building converged at the top of the crown. Only one corner of the building had flats that were not on fire. A handful of residents remained in the tower alive and unable to escape – the blaze closing in.

Assistant commissioner Andy Roe believed the building was still structurally sound enough to keep sending firefighters up on rescue missions. More extended duration breathing apparatuses arrived. But the conditions in the tower remained perilous – dark and hot, with firefighters asked to try and effect rescues, without anything to tackle the fire and radios prone to cutting out. A decision was made not to commit firefighters above the eleventh floor.

*

At this point, the eleventh floor was incredibly hot. One crew who reached it recorded temperatures of above 1,000°C on their thermal imaging cameras. 'When we got to the eleventh floor I knew that pretty much every flat was on fire, I knew this because when I put my thermal image camera really close to my face the screen was all white,' he recalls. He could feel the heat through his protective hood. The team had no water or hoses to fight the fire. 'If we had breached one of the compartments we could have set the whole floor off on fire. I made a decision that this was a dangerous place to be, I knew that if it went wrong it was going to kill me and my crew, I felt a huge sense of responsibility... I knew that we had to get off that floor.'

But behind these doors, Natasha and her family were still trapped. At around 4.15 a.m., the smoke was building up in her bedroom to such an extent that they fled to her daughter's room. But when they closed the door, they realised they were trapped: the door handle was missing and they could not open it. They were able to prise it open with a pair of scissors and escape, but it was now clear: they had flee the flat or stay there and die.

At 4.27 a.m., Natasha phoned her sister Denise who was outside the tower and was told firefighters were coming up to them. Her partner stood by the door on the eleventh floor shouting 'hello' and suddenly, Denise, standing outside, heard his voice. It was being transmitted through the radio of a firefighter standing next to her outside the tower. She knew the crew searching for Natasha must be nearby.

Natasha's partner opened the door onto the smoke-logged landing. She heard a firefighter shouting: 'Can you see the torch, walk towards the torch.' They guided them to the stairwell and down the stairs and out of the tower. They left the building at 4.47 a.m.

*

At 5 a.m., it fell eerily quiet in the control room in Stratford as the phones that had been ringing incessantly all night fell silent. There was a stunned silence as the terrible reality of what had happened began to settle in.

Crews were still being deployed up the tower on search and rescue missions, although many of the flats they entered were by this stage empty or wholly on fire. There were still fears about the collapse of the building and crews were told to look out for any misplaced columns as they moved up and down the tower. The structural engineer was now on-site, however, and the firefighters continued with the rescue operation.

One crew made it to Floor 10, where they rescued the occupant – Antonio Roncolato – who had been trapped for hours in his flat. At 6 a.m., he heard a knock on the door and opened it and two firefighters came into the flat, closing the door behind them. He put on his son's swimming goggles, pulled a wet towel over his head and was helped down the stairs and out of the building.

The firefighters running the operation were trying to gain a picture of who was and wasn't accounted for. They thought 171 people had at one stage or another been trapped in the building and called for help, altogether comprising 38 children and 133 adults from 45 flats – but it was not clear who had been saved and who was left. The firefighters thought that 115 people were unaccounted for, although it was difficult to confirm the exact number.

As such, the senior fire brigade officers wanted a list of residents from the council or the TMO. But this was not forthcoming. The first request is thought to have been made at close to 5 a.m. By 5.50 a.m. – at the next tactical meeting of the senior members of the emergency services – it still had not been provided. They also wanted up-to-date plans of the block. This became a source of major frustration for the firefighters. Tempers were frayed. At 7.16 a.m., a police officer's body camera records Andy Roe expressing this frustration to the liaison officer. 'The fact that you've not been able to get me a set of plans is a major deficiency and will be highlighted,' he said.[1]

A liaison officer for RBKC had first asked Robert Black, the chief executive of the TMO, for the list at 4.50 a.m. At 6.24 a.m., Mr Black was emailed a list of occupants correct as of 30 May 2017. Plans of the building were sent to him at 6.14 a.m. But he did not forward these plans to firefighters until 7.35 a.m. And the list of residents was not sent until 7.56 a.m. He later claimed he did not realise they wanted it. In the meantime though, Mr Black had sent an email around his team discussing the media strategy. 'We need to pull some of this together pretty fast in terms of Health and Safety compliance,' he wrote. 'We need all the information about the refurbishment as this will be a primary focus.'[2]

While all of this was going on, the world was waking up to what had happened overnight at Grenfell Tower. Television news channels around the world were broadcasting pictures of the ruined tower – the charred walls and the fires still burning inside the flats.

These cameras began to train on one floor in particular – the eleventh. Amid the plumes of smoke and fire still burning on the cladding, the outline of a man could be made out at a window waving a white towel.

16

A BLIND SPOT

It is hard to overstate the impact of the failure to evacuate
Grenfell Tower. There have been many cladding fires
around the world, but most of them have had no fatalities,
with all the residents simply walking out of the building
unharmed. This includes a string of enormous fires in the
UAE, and one in Grozny, Chechnya, in 2012 when – as at
Grenfell – fire spread right around the building and completely destroyed it.

The stay put policy is at the heart of the Grenfell Tower
story. The morning after the fire many news bulletins led
with reports that residents had not heard any fire alarms.
Really, this shouldn't have been a surprise. Grenfell didn't
have any. Neither does almost any other residential block
in the country.*

From a technical perspective, stay put is founded
on the idea of compartmentation. With effective

---

* Communal fire alarms are distinct from individual smoke alarms,
which are a legal requirement in all residential properties including
flats of any height. However, they are not interconnected and cannot
send an alarm throughout the building.

compartmentation, a building will contain the spread of fire within a single unit long enough to allow the fire-fighters to extinguish it while all the other residents go about their business unaware that anything untoward is even happening. In fire services and the housing world, there is a firm belief that UK buildings are built to standards that effectively guarantee the success of this policy.

This certainty is based on ideas developed in the 1960s, and has its roots in our history, which has seen our fire codes prioritise stopping the spread of flame from building to building, since the Great Fire of London. As such, as we started to build upwards, building codes continued to try and prevent fire from spreading from property to property – and also from one flat in a single block to another.[1]

There is nothing wrong with this idea in principle. If followed and maintained, it makes for solid, sturdy buildings in which fires are often minor incidents that do not require mass evacuation. The problem is that our faith in the policy has grown to the point where we have no Plan B for what to do when it goes wrong.

In reality, the idea of perfect compartmentation was always a fiction. Fire research from 1960 – before the idea of stay put was even born – demonstrated that fire could spread up the outside of a building by breaking out of a window and licking up to the window above without even the assistance of combustible materials and even if windows were spaced quite a long way apart.[2] Moreover, the very act of fighting a fire will breach the compartment. Once firefighters have knocked down the door to the burning flat and propped open the door to the staircase to run their hoses through, smoke will start to spread through a building.

But, before Grenfell, any fears about stay put were allayed with the regular use of statistics which apparently demonstrated how well the policy was working. For example, the National Fire Chiefs Council emphasised after Grenfell that of 57,000 fires in high-rises between 2010 and 2017, only 216 required the evacuation of more than five residents.[3]

But scratch beneath the surface and this data is far less reassuring than it sounds. As the expert witness Professor José Torero told the inquiry, fire service evacuation data is misleading. It does not tell you about the incidents where residents escaped themselves before the firefighters arrived, or where they should have been evacuated but weren't. And the low percentage of fires spreading beyond two floors is less comforting when you realise that it is still happening once per week. Detailed analysis of fire data by Dr Stuart Hodkinson, high-rise safety expert Phil Murphy and researcher Andy Turner painted a different picture. They found 1,847 fires in the 2010s which spread outside the flat where they started. Moreover, they found a worrying jump in the fatality rate in these fires: 29.6% resulted in a death compared to 15.6% where the fire did not spread. They also revealed that firefighter response times were far more likely to be delayed in blocks of flats, as firefighters struggled to access the building and to locate the affected flat. This gave more time for the fire to spread.[4]

Professor Torero was scathing about the comfort drawn from statistics on this point. He said it pointed to 'incompetence at all levels'. Even if the probability of an out-of-control fire in a high-rise was low, risk is not just about probability. It's also about consequence. '[If] the consequences are massive, you cannot ignore the potential of the consequences even if the probability is small,' he said.[5]

One reason to play down the risk is a misguided fear of a panicked evacuation, which is particularly prevalent within the fire service. This is not actually supported by the evidence. As one lawyer for the bereaved and survivors, Danny Friedman QC, told the inquiry: 'Those who have studied crowd psychology during disasters, including during fires and explosions such as the Summerland resort, the King's Cross fire and the collapse of the World Trade Center, have established that, in moments of crisis, people tend to act in an unusually collaborative and bonded way, even with strangers, but especially when they are in familiar surroundings and with people who are known to them.'[6] The evidence from the night of the Grenfell Tower fire only adds weight to this belief. Those who did escape mostly did so calmly, collaboratively and, particularly in the early stages before the corridors filled with smoke, without help from firefighters.

There were also practical and financial concerns. Stay put is an enormously convenient policy and were we to acknowledge it needed to change, many implications would follow. We would need to stop allowing blocks to be built with a single staircase – which would trim the profitability of housing developments considerably. We would need fire alarms, which housing providers would need to install and maintain. In council housing, these would have to be paid for by the state. Perhaps, if we accepted that blazes could spread outside the compartment of origin, we'd also need sprinklers to put them out. And we'd have to figure out what to do to rehouse the hundreds of thousands of disabled residents who would not be able to escape a fire, or put in place provisions to allow them to do so. All told, it is simply easier and cheaper to stick with stay put – and to close eyes and ears to the evidence that our total reliance on it may be misplaced.

A DOUBLING DOWN ON THE STATUS QUO

Until the Lakanal House fire in 2009, there were no major disasters big enough to grab public attention and result in major scrutiny of 'stay put'. But after that terrible summer's afternoon in south London, it should have been obvious to anyone who considered the question that something needed to change. This fact became inescapable after March 2013, when the jurors' verdict at the inquest decided all six victims could have escaped if they or their parents had been told to get out of the building by call handlers and were aware of the escape routes.*

---

* Narrative verdicts said Ms Hickman would have been able to escape up until 4.40 p.m., having first phoned 999 at 4.21 p.m., and those trapped in the bathroom could have escaped up until 5.15 p.m. Verdicts accessed via https://beta.lambeth.gov.uk/about-council/transparency-open-data/lakanal-house-coroner-inquest

Now was the opportunity for change. Stay put may work for many fires in many flats – but it didn't suffice. We needed to plan for the next time a fire got out of control in a high-rise block. But no plans were made. The government chose to entrench the status quo instead.

In the aftermath of Lakanal, two years before the jurors delivered their verdicts, the housing sector asked the government for guidance on what it should do about fire risks in blocks of flats. In 2011, a renowned fire safety consultancy C.S. Todd & Associates was commissioned to write a fresh guide for fire safety in high-rises which would help social landlords respond to Lakanal. It was written with the close involvement of Eric Pickles' department and was published with resounding support from the fire and housing sectors and a range of expert voices.

The guidance was a ringing endorsement of stay put. The principle was 'undoubtedly successful' it said: 'In 2009–2010, of over 8,000 fires in these blocks, only 22 fires necessitated evacuation of more than five people with the assistance of the fire and rescue service.' It did say fires requiring evacuation happened, but added: 'Fortunately, these fires are rare. They are usually the fault of failings in the construction.'[7] The guidance was making precisely the mistake Professor Torero would later identify. A low probability event with calamitous consequences should still be something planned for carefully.

The guide also expressly advised against fire alarms. 'General needs blocks of flats [as opposed to specialist housing for disabled residents] will not normally require a communal fire alarm system,' it said, adding that they 'should not be installed unless it can be demonstrated that there is no other practicable way of ensuring an adequate level of safety'.[8] But there was another big question and it would go on to become deeply controversial after Grenfell. What should be done to ensure residents with disabilities can escape a fire?

AN OUTDATED VIEWPOINT WHICH IS HIGHLY DISCRIMINATIVE

Before the LGA guide was published in 2011, the legal position on the evacuation of disabled residents was clear. A wealth of laws and guidance documents set out the position: measures must be put in place to ensure they could leave the building in an emergency without reliance on firefighters.* This was a simple question of equality, and is applied without question as a standard part of risk assessments of workplaces – through a process known as 'Personal Emergency Evacuation Plans' (PEEPs).

But in blocks of flats, this law was not being implemented. Colin Todd, the director of C.S. Todd & Associates, when he gave evidence to the Grenfell Tower inquiry, explained that the housing sector had interpreted the law as only applying to workplaces, despite this caveat not actually appearing in the legislation itself. Asked why this was the case, Mr Todd said 'we trusted stay put basically.'[9]

---

* Guidance documents overtly requiring PEEPs include PAS 79 and British Standard 9991.

This position would be codified by Mr Todd when he – along with colleagues – wrote the 2011 guidance document. This encouraged reliance on stay put for disabled residents, telling social landlords it was 'usually unrealistic' to plan for or 'to have in place special arrangements' for disabled residents. No disability experts or representative groups of disabled people were consulted in preparing the guidance. Instead, the advice was based on the 'practical concerns of landlords' that providing evacuation plans would be too difficult.[10]

This raised concern. The Chief Fire Officers Association criticised the guidance when it was presented to them in draft form, saying that to 'ignore and eliminate advice on disabled access and evacuation is a fundamental error'.[11]

After the document was published, a consultant, Elspeth Grant, also raised concerns, warning that the guidance 'reflects an outdated viewpoint which is highly discriminative and not in line with UK legislation relating to equality or fire safety' and could result in an 'unnecessary tragedy' if it resulted in disabled residents being left without escape plans. The response to her letter was drafted by C.S. Todd & Associates. It insisted that it was not 'reasonable and practical' to recommend evacuation plans for disabled people 'by way of default in all blocks of flats'. To do so 'would place a significant burden on those managing blocks of flats to continuously update the information'.[12]

Such 'practical concerns' often arise when PEEPs are discussed. What should be done, critics of the idea ask, about residents who cannot escape unaided and do not have live-in support from carers or relatives? General needs blocks are distinct from care homes and offices which have staff present to offer assistance. If the housing sector were required to put staff on-site in every block twenty-four hours a day to evacuate residents the costs would be extraordinary.

But this typical response overlooks what a PEEP actually is. Disabilities come in various shapes and sizes. For a deaf resident, for example, it might be a vibrating pillow synced to the smoke alarm. For an autistic resident, it might be a sensitively discussed and clear plan for how and when to leave the building quickly. For someone with severe arthritis, it might be low pressure door handles. For a mother and baby it might be a smoke hood. None of this requires the presence of staff.

For those who do require assistance to get out of the building, the question is undoubtedly more complex, but not necessarily as difficult as is sometimes made out. Many disabled residents of blocks of flats do have someone living with them or nearby who helps out. This person could be trained to help them into an evacuation chair and out of the building, in the same way that office workers with disabilities are given 'evacuation buddies'. People rarely leave a disabled loved one to die in a fire. If there is no means of evacuating in a fire, the likelihood is both will die. This is what happened at Grenfell Tower in the tragic case of Sakina and Fatimah and others.

For those who really do have no one, they could be offered priority status for rehousing to a ground floor property. Such houses are in short supply, but over time rehousing could take place – especially if our new build policy was targeted more towards homes suitable for those with disabilities. Many of these residents may seize the chance to move to a more appropriate property where they would not be imprisoned every time the lifts broke down. Others may decline, and elect to accept the risks of staying in a block of flats they love. But at least this would then be their choice, rather than circumstances – like Sakina – they had been forced to accept.

The above would still cost money and time. But in a country as well-resourced as ours, it is something we can afford. It would likely have many unplanned benefits, in forcing us to reach out and offer support to disabled residents who are too often forgotten in high-rise flats without support and save lives even in smaller fires. 'If the flat of the disabled person is on fire or the one next door is, they have to be able to move away from the fire. And if they have a problem moving … they can't. After 20 minutes in the near vicinity of a fire, the chance of a fatality is extremely high,' Elspeth Grant told me in 2021.[13]

What's lacking is will, both at a senior political level and from the housing sector that has always lobbied against PEEPs, fearing the administrative and legal burden that would come with providing them. Ms Grant says it is 'endemic from top down to bottom up' in the fire and housing sectors that the policy is not properly implemented.

As it was, the 2011 guidance gave backing to the position that such plans were not needed. And this guidance was followed. For Mr Todd's part, he remains a vociferous advocate of stay put, which he says is 'advantageous' to disabled people. If a block is built in a way which means evacuation is not required, that means a disabled resident does not have to undergo the risky process of evacuating from it during a fire. But, as is the case with all our reliance on stay put, that line of thinking requires us to ignore the difficult question: what happens when it all goes wrong?

A LANDMARK ACT OF DISCRIMINATION

This is what happened in west London. It seems that the TMO originally intended to prepare PEEPs. Board papers from July 2010 refer to a plan to 'identify residents with special needs and work with them to establish a specific PEEP to ensure their safety is protected'.[14] And chief executive Robert Black emailed internally to say the organisation planned to prepare 'generic PEEPs' which could then be used to produce tailored plans for individual residents in the same year.

But this work simply drifted away. While data was gathered, it was held across two separate systems and no work was ever done to understand how those with disabilities would escape in a fire. Why not? When asked about this at the inquiry, the TMO witnesses referred to the amount of resources required to do it and the practical difficulties of evacuating residents without any staff on-site to assist. They also said there was a 'conscious choice' to follow the guidance produced in 2011.

As it turned out, at Grenfell Tower thirty-seven of the occupants present when the fire broke out had a disability which meant they could not evacuate unaided. By the next day, fifteen of them had died – unable to escape the burning building with no plans in place for them to do so. The fire would go on to be branded 'a landmark act of discrimination' against the disabled and vulnerable by lawyers acting for the bereaved and survivors[15]

A BLIND SPOT

If the housing sector was relying on the LGA guide, and its ringing endorsement of stay put, what about the fire services? The coroner investigating Lakanal House, Frances Kirkham, wanted to see change. In her letter, she called on Eric Pickles to publish 'consolidated national guidance' on 'the stay put principle and its interaction with the get out, stay out policy'.[16] Mr Pickles responded with a promise to provide 'advice to [fire service] incident commanders to inform decisions on evacuation should it become clear in a fire that the stay put principle is no longer tenable'.[17]

This involved revising national guidance for fire services contained in a document called 'Generic Risk Assessment 3.2'. The LFB had, in fact, already been appointed by central government to lead a redraft of this document to include the lessons from Lakanal.

But the organisation had shown a strange reluctance to include anything on reversing stay put. A draft which went out for consultation in February 2012 did not mention needing to reverse the policy. Other officers who read the draft commented that something should be added specifically dealing with evacuation; their remarks were not incorporated. Asked to account for this, Peter Cowup, the senior officer who led the drafting process, said the brigade had a 'blind spot' regarding evacuation.[18]

But officials at DCLG were worried. One wrote in an internal email in October 2013 that it 'does not clearly meet our commitments to the coroner'. Despite prompting from officials, Mr Cowup still did not amend his draft. Ultimately, officials took things into their own hands and unilaterally added a passage to Mr Cowup's draft before the document was published in February 2014.[19]

This said fire authorities should have 'contingency plans' including 'an operational plan... in the event the "Stay Put" policy becomes untenable.' But beyond this, nothing was written on how to do it. While LFB's internal guidance reflected the need to carry out a 'full or partial evacuation' of a high-rise building, it never developed a training programme for its incident commanders on when and how to do this.

Such a plan is possible. Dr Michael Reick, a regional fire chief in Germany and former leader of the fire research laboratory at the University of Stuttgart, says: 'There is always a chance that buildings fail catastrophically.' He explains German firefighters are trained to use fans, curtains and smoke hoods to help facilitate evacuation, even when compartmentation has failed.

But Dr Reick is clear that this was not the fault of those present on the night. The plan should have been developed years in advance. 'People in the UK blame the fire department for not having a Plan B [at Grenfell]. I can understand this. But this Plan B should have been prepared by fire engineers years before. These preparations did not happen,' he says.[20]

And so those who took command at Grenfell Tower on the night of 14 June 2017 were left to improvise. This was always bound to fail. Michael Dowden and those who followed him were given no plan. The failure to provide one belongs to the whole British state, from the fire services up to central government, who chose blind faith in stay put over the steps needed to prevent an entirely foreseeable disaster.

AN ARTICLE OF FAITH

Sadly, Sir Martin's stark finding that stay put was applied as an 'article of faith' and more lives could have been saved if it was reversed earlier has not prompted much of a change. The policy retains its power in the housing and fire sectors and within government. I remember attending a housing industry event, where someone suggested it was time for a 'debate' about stay put, almost two and a half years after Grenfell. Another member of the panel disagreed angrily. 'We don't need a debate,' he said. 'We need to shout it from the rooftops. Stay put works.' When I took my children to a local fire station open day in October 2021, we were handed a leaflet telling us what to do if there was a fire in a block of flats. 'Stay put and call 999,' it said, adding that it 'may be safer to stay in your flat or maisonette' until fire services arrive, even if there was smoke or fire inside the flat.[21]

The truth is buildings need what fire engineers refer to as layers of protection: sprinklers to control the fire and fire alarms and second staircases to provide for evacuation if necessary. Most building codes around the world provide back up plans. Even if the advice is to stay put initially, a 'phased' evacuation can take place if needed. The inquiry's expert Professor Jose Torero called the UK a 'global outlier' due to its total reliance on stay put.[22]

'Stay put is fine as a Plan A,' Professor Ed Galea, an internationally respected expert on evacuation told me in 2022. 'But you have got to have a Plan B, and then a Plan C. Plan B should be the residents of the building evacuating themselves and then Plan C can be the fire service rescuing those still trapped.'[23]

Before Grenfell, we had no Plan B or Plan C. All our chips were stacked on 'stay put', due to a hubristic conviction that UK buildings were too good to fail. This gamble was lost, and those who died at Grenfell were the ones who paid the price.

17

8 A.M.

Elpidio Bonifacio had lived on the eleventh floor of Grenfell Tower for thirty-six years. His wife Rosita was out on the night of the fire and he was home alone.

Elpidio had been registered as blind for fifteen years. He could not see things clearly or recognise objects. Since the refurbishment, he had found moving around the tower more difficult, because the familiar layout of the building had changed with the addition of three new floors of housing at the lower levels. When the lifts were not working – which happened frequently – he got lost between floors and could not find the exit to the building.

He had woken from his sleep earlier in the night to a phone call from his wife frantically warning him about the fire in the tower. He opened the door to their flat and could smell the smoke. It was a strong smell and he could feel the temperature was warm, hotter than usual.

His son and daughter-in-law had phoned the fire brigade on his behalf to tell them where he was. He packed a rucksack with passports and other important documents and waited for them to arrive.

But he had waited and waited and they had not come. Despite his limited vision he could see burning objects falling past his window. From his sitting room, he had been able to call his wife and son who were also on the phone to the emergency services passing on his location. But eventually, he believed the fire had reached the living room so he retreated to his bedroom, shut the door and began waving his white towel from the open window. This was the image that was now being watched by millions on live news footage around the world.

He began to feel the water from the firefighters' hoses outside the building. It was so cold that he needed to step back into the room to keep warm, but he had to keep going back to the window for fresh air. The room was filling up with smoke.

Elpidio considered jumping. He tried to drag the mattress from the bed and throw it out to cushion his fall, but it was too heavy so he gave up. He began to feel hopeless. The flat was becoming hotter and he thought he was about to die. He picked up a handful of his medication and prepared to take it, to at least numb the pain. But with suicide prohibited by his Catholic faith, he decided to leave his fate up to God. He heard the mirror in the sitting room shatter and the crackling of flame outside the door. The smoke alarms throughout the flat began to blare. 'Lord into your hands I commend my spirit,' he prayed.

*

Watch manager Andrew McKay and his team were advancing up the tower wearing extended duration breathing apparatus at around 7.30 a.m. to fight the fires which were still burning on the upper floors. As they climbed the stairs, the ground was littered with hoses and water was flowing down the steps. On the eleventh floor, they entered the lobby and began to search the flats. The first was completely burnt out and they left. The next was also empty. The firefighters could see the towels that had been put down on the floor by the residents in a bid to block out the smoke. It is likely to have been the one Natasha Elcock and her family had fled only hours before.

The next flat was very hot, and there were two fires burning in the hallway. One of the firefighters began trying to extinguish them with the hose. Suddenly a bedroom door opened and an elderly man appeared.

'They were big men, they looked like giants to me,' Elpidio recalled in his witness statement. 'They said, "Come on, relax, we will take you down." The Lord had answered my prayers.' Elpidio was carried out of the building at 8.07 a.m. He was the last resident to leave the tower alive.

'SPACE SHUTTLES ON THE SHARD'

The LFB is a much-loved institution. The idea that it was to blame for what happened at Grenfell Tower prompted much outrage when the inquiry's first phase report criticised its actions on the night. How could the firefighters who risked their lives be taking the blame which should be directly aimed at corporations and governments?

But the uncomfortable truth is the LFB has difficult questions to answer. While the scale of the fire at Grenfell Tower shocked everyone, the brigade should have had a plan for a major, out-of-control fire in a high-rise block. That it did not was a result of hubris and a failure to act on the need for change.

Take the issue of call handling. The inquest into the six deaths at Lakanal House clearly identified the problems later exposed at Grenfell. The jurors said operators had 'a clear expectation that persons trapped would be rescued by firefighters' and 'their advice… relied heavily on this assumption'. Training was said to have 'failed to encourage active listening' or 'encourage operators to react to dynamic or unique situations'. Training documents were also branded 'contradictory and inconsistent' particularly in relation to the issue of advising a caller on staying put or getting out during a fire.[1]

But the coroner did not call for the LFB to make changes, noting the 'steps already taken by the brigade'.[2] In fact, the steps being taken were partial and poorly implemented. From as early as 2010, an internal LFB audit found 'adequate long-term planning' of training for control room staff was 'not being carried out'. The content of the training was also flagged as a problem. 'I feel if we were to take this PowerPoint to the [internal LFB board scrutinising the response to Lakanal], we would be severely criticised. It fails to cover some issues we have already informed the board are dealt with,' wrote one control room manager in 2011.[3]

Even after the inquest, the training was cut to four hours and the role play element was dropped. Staff carried it out in a room adjoining the control room and could be called back to their desks if pressure demanded. The training should have been annual for all staff, but in 2013 figures show only 28% taking it, with 29% in 2014 and none at all in 2015. This was because of a lack of resource, and major problems with an IT upgrade soaking up staff time. The reality was that eight years after Lakanal, the LFB's call handlers were hardly any better prepared, and went on to repeat the mistakes on an even larger scale.

The training of incident commanders – in particular about when to revoke stay put also fell by the wayside. Despite the brigade telling its internal monitoring body the work to develop this training was complete from autumn 2013, it had barely begun. Problems with a privatised training provider, Babcock International, meant it was not fully rolled out at the time of Grenfell, and did not include the crucial point about revoking stay put. LFB's frontline staff, from the control room to the incident ground, were set up to fail by a lack of training on what to do if stay put failed.

And the brigade, at least at the top, knew stay put could fail. In March 2010, for example, the LFB responded to a Greater London Authority consultation with a warning that modern cladding materials could 'provide a route for fire to spread by bypassing cavity barriers or fire stopping'.

The organisation wrote to the Home Office in December 2012 warning that a fire in a tall building with combustible material on the outside has 'potential to affect multiply [sic] storeys simultaneously, thus making fire-fighting more difficult'. After the Shepherd's Court fire in 2016, the brigade considered warning other building owners in London about the danger of combustible materials on their facades. 'This could be the proverbial "cat out of the bag" on this issue,' one senior staff member wrote.

In fact, its fire safety experts even developed a PowerPoint presentation setting out the risks of cladding fires elsewhere in the world. 'Facade fires and uncontrolled spread of fire across the outside of tall buildings is a significant threat to the effectiveness of the many systems within the building,' wrote one member of the technical team, in notes for a presentation intended to be delivered in June 2017 – just days after Grenfell. But despite all this knowledge the experts simply never planned for the evacuation of a tall building, seeming to hold some sort of British exceptionalist belief that it simply wouldn't happen here.

When she appeared at the inquiry after the fire, then commissioner of the LFB Dany Cotton angered survivors by comparing the brigade's failure to plan for a fire like Grenfell to a failure to plan for 'a space shuttle landing on the Shard [London's tallest building]'.[4] But this was spectacularly disingenuous. The LFB knew a cladding fire could come. It failed to prepare.*

'WHAT [BEREAVED AND SURVIVORS] WANTED WAS NOT HEROES, BUT WELL-TRAINED PROFESSIONALS WORKING TO A WELL-STRUCTURED PLAN'

The LFB was repeatedly described at the inquiry as conservative, cumbersome and resistant to change.

Around the country, other fire services adopted a more nuanced process to decision making during incidents – informed by research on the psychology of incident commanders when under pressure. The senior firefighter who helped develop this model was seconded to the LFB before Grenfell, but was told the brigade would not adopt it. Discussing this at the inquiry, she said there was a culture

---

* In its opening statement, the London Fire Brigade said that while international evidence offered 'much to learn about fire behaviour', those fires mentioned were not like the Grenfell Tower fire, and did account for the fact that UK buildings were not designed for rescue and firefighting on multiple floors.

in the LFB that was 'very conservative' and 'there is great comfort in what is familiar'.[5]

Lawyers for the bereaved and survivors pointed to a management culture that saw top rank officers work their way up from operational firefighting, without any serious management training. They then promoted those coming up behind them. This, the lawyers said, fostered a culture where the accepted way of doing things was seen as gospel and senior officers were sceptical of new ideas. One lawyer, Danny Friedman QC, described the culture as follows: 'They excelled in engaging with fire, muscle–memorising the standard techniques, fitting into watch culture and functioning as hands-on charismatic leaders… They represent an ideal of courage and community of service associated with the nineteenth-century origins of firefighting.'[6]

In a landmark study of the psychology of the fire service, the sociologist (and former firefighter) Dave Baigent suggested the strong dominance of a 'highly male-gendered need to get into the fire'.[7] This heroic instinct to tackle and extinguish the blaze is the way UK fire services are set up, and what the stay put policy caters to. In Milan in August 2021, the city's fire service responded to a huge cladding fire by simply letting it burn, and ensuring everyone was out of the building safely. While this fire was not the same as Grenfell (the building was partially occupied and the fire occurred in the afternoon not the middle of the night), it demonstrates the difference in philosophy. Nobody died in Milan, despite the building being lost. But acting as the enablers of an evacuation does not fit with the way the British fire service sees its role. 'What [bereaved and survivors] wanted was not heroes, but well-trained professionals working to a well-structured plan,' added Danny Friedman QC.

Darker sides to this culture could occasionally be glimpsed during the inquiry evidence. The inquiry saw images, used in staff training, of a bare-chested male firefighter carrying a woman in lingerie out of a blaze. Current commissioner Andy Roe recalled the openly racist comments he heard during his time on the watch – telling one anecdote which involved a firefighter saying 'Pakis breed like rabbits' after leaving the scene of a blaze at the home of an Asian family.* He has said he would not be comfortable with his mixed-race daughter joining the brigade. Imran Khan QC, representing one group of bereaved and survivors said, considering these issues, the brigade should 'at the very least consider whether it is capable of delivering an appropriate service to a population as diverse as that which exists in London'.[8]

'GET STUFFED'

There is a broader story to the failure of the fire service. This story begins in 2004. New government legislation

---

* Inquiry transcript, 30 November 2021. Mr Roe said he believed the example 'exemplifies the reality' of watch culture, but added that he does not think 'race or misogyny plays any part in our response'. He had launched a review of race and gender discrimination that had not reported at the time of writing.

abolished national standards and scrapped a national body, the Central Fire Brigades Advisory Council. Central government control over fire services all but disappeared overnight and in its place came a doctrine of localism.

Big policy decisions were delegated down from elected national governments to the individual forces: forty-six different authorities working to their own scripts, defined by their own local priorities and local leadership. There was no higher authority which might have held the LFB to account.

And then the devastating funding cuts of the 2010s hurtled in. While more firefighters or more fire engines are unlikely to have made much difference to the outcome at Grenfell, cuts undoubtedly contributed to the failure to train and plan in the years before. And these cuts were hefty.

In 2013, Sir Ken Knight, the senior government advisor and former firefighter who had also rejected the need for retrofitting sprinklers, published a review of fire and rescue services for the coalition government. Pointing out that call-outs to fires had reduced 40% in the last decade while spending had remained consistent, he said £200m in 'savings' could be found by fire authorities through the implementation of 'efficiencies'.[9] This meant cuts – primarily to staff. According to the National Audit Office, by 2017, fire authorities had lost a quarter of their staff. In 2010, they had 41,632 full-time equivalent employees. By 2017 the number was down to 32,761.[10]

These cuts were implemented in London by the mayor at the time: Boris Johnson. He pushed them through despite their rejection by the LFB's governing body. In the eight years he was mayor of London, the brigade experienced cuts totalling £100m. It cut 552 firefighters, 27 fire engines and closed 10 fire stations. Crucially, the cut to jobs particularly affected support staff – with 324 posts removed, 29% of the total.[11] Senior officers' roles were also cut – giving the London Fire Brigade a much thinner ratio of support staff and senior officers to frontline firefighters compared to other forces. It is hard to believe that this did not contribute to the failures to plan and prepare before Grenfell. It has also reduced London's resilience more generally, which should scare us given the hotter world we are facing. In July 2022, when record temperatures caused major fires around the capital, the thinly resourced brigade struggled to cope.

In a debate about the mayor's plans in September 2013, a Labour politician accused him of lying about the impact. Johnson told him to 'get stuffed'.[12]

AFTER THE FIRE

As day broke in west London, a severe humanitarian tragedy was unfolding. All those who had fled the tower were destitute. All their possessions were destroyed, from birth certificates to toothbrushes. Meanwhile, family members and friends had congregated at the base of tower, desperate for news of their loved ones.

Gradually the crowd had been moved back from the tower by police, into the leisure centre. 'The police were shouting at us to move back. It felt like "public order" superseded caring for the victims,' Eddie Daffarn said. An ambulance driver shouted at him to 'go home' – not knowing that his home had been destroyed. He recalls a police officer swearing at him. 'The only real help we received at this time was from local people offering tea, water and comfort. I don't remember anyone from the emergency services or local council coming to see if we were okay,' he said. 'We were just abandoned by the side of the road, in distress.'

The local council, RBKC, was legally required to respond to the disaster by setting up 'rest centres' for those displaced and arranging emergency housing for them to move into. The British Red Cross were on the ground, and some local churches, youth centres and mosques had opened their doors. But the council were not there. Its staff were not deployed to the scene, and at the town hall the emergency response team struggled to find the keys needed to open up their major incident room. As a result, the situation on the ground was a mess: people did not know where to go or who to ask for information.

Karim Mussilhy, who travelled to the tower searching for his uncle Hesham, described the scene as he arrived at 7.20 a.m. 'I couldn't believe what I was looking at, but it was like something from a horror film or disaster film. It was just crazy and the smell – I can still remember the smell of this burning plastic – I've never smelled anything like this before in my life. I can still remember the smell to this day,' he said.[1]

He went around the makeshift rest centres, but was denied entry because he was not a survivor of the fire. No one from the council or the TMO was visible. He said although police were present after the fire, it seemed like they were 'ready for a fight' and told anyone who wasn't supposed to be there to 'move along'.

He said he first realised his uncle had not made it out when he saw firefighters' T-shirts placed as memorials at the base of the tower. Firefighters had written notes on them, one of which said 'to all those on the 21st floor and above we are sorry we couldn't get to you'.

Karim's uncle lived and died in Flat 204 on the twenty-third floor of the tower. He had health problems and struggled with mobility, but – like with so many others – the council still placed him near the top of the tower without assessing his ability to escape in an emergency.

Nabil Choucair told the inquiry of his panicked hunt for six family members who lived in the tower. 'It was very unclear; it was so unorganised. You would think in a situation, in an emergency catastrophe, there is some form of plan, some sort of organisation but... everything was falling apart, and it was just so unorganised. It was so unhelpful. It was like we were trying forever, but with no help, or no sense of help or [being told] exactly what to do,' he said.[2]

His brother Hisam went to eleven hospitals looking for news of their family. 'It was like the inside of your gut was being ripped up [because of] the lack of communication, the lack of updates,' he said.

It would be weeks before it was finally, officially confirmed to the brothers that their six relatives had all died in the blaze. Other families would receive the news they were dreading via word of mouth in the melee outside the tower. Their screams of shock and grief attracted press photographers, who plastered the images of what should have been their most private moments across the next day's newspapers. Others still would have no official confirmation until August, or learn of the death of their children from news reports before it had been given to them in person. Some would never get official word at all.[3] The lack of information meant an official death toll was hard to define. But police also evidently feared a release of a much higher number would spark unrest. In a risk assessment produced four days after the fire, the local police branch wrote: 'There is an expectation that the death toll from the fire could rise substantially. And with the cause unknown, any subsequent disclosure would have the impact of community tensions, especially when the majority of those affected are believed to be coming from a Muslim cultural background.' The document was described as 'Islamophobic' and 'racist' by lawyers for the community.*

Nicholas Burton, along with a number of other survivors, had been taken to the leisure centre sports hall after he and Pily fled the tower. It was a chaotic environment: paramedics attending to shell-shocked, soot-covered residents, many laid out on gym mats. They were examined and treated before being carried to ambulances and driven to hospitals around London. Natasha Elcock was also taken to hospital, following her harrowing descent down the smoke-filled stairwell.

In order to ensure everyone had support, the council needed a register of who was in need of help. People gave their names as they arrived at rest centres. But the council had no system for collating this information. Boxes of paper forms with names and details of those impacted were simply left scattered around the churches and halls where survivors were congregating.

This was a problem. Without clear details of who was affected, reaching them to provide help in the days that followed became impossible. Residents were offered hotels all round west London and then effectively abandoned.

---

* Inquiry transcript, 27 June 2022. The Met Police said the document also referred to working with the local Muslim community and 'strongly refute' any allegations of Islamophobia.

Whole families were placed in single rooms, recently bereaved relatives were left with no support, parents were left without facilities to sterilise bottles for their babies, residents, including those with disabilities, were put in upper floors – traumatising after the blaze. No support was available to those with disorientated family members with conditions such as dementia. Some stayed in these hotels with no contact from the authorities for days. Others were evicted at a moment's notice because the bookings had been allowed to expire.

Then there were 845 people who lived in low-rise blocks connected to the tower. They had been evacuated during the fire, and their homes were now behind the police cordon and in a state of disrepair: doors had been broken in by police during the evacuation, many were flooded with water which had been sprayed at the burning tower and all of them were without heating, hot water and cooking facilities: the communal boiler at the base of the tower, which served all the homes, was out of action.

But they were not offered rehousing – with the council deciding that doing so was simply too difficult. Instead, they either found space with friends and family or slept on mattresses in the leisure centre on the estate. Many – unaware that this was an option – slept rough. The next day they would be told to return home to houses without power or heating or water. Their children would play in the dust and debris of the fire that had killed their classmates.[4]

The police, meanwhile, treated the community aggressively. Relatives searching for their family members were shouted at and threatened with arrest. Bereaved family members recall being treated like 'criminals'. There are reports senior police officers expressly said they feared 'another Duggan' – in reference to the riots that followed the shooting of Mark Duggan in north London in 2011. Anti-terror operations were put into place and witnesses recall armed police patrolling the streets around the tower as families searched for their lost loved ones.*

'I cannot help but feel that had our community lived in a different part of the borough, on the more affluent side, had we been from a different class, had we been less ethnic, the response in the aftermath would have been immediate,' Hanan Wahabi, who escaped the tower but lost her brother and his family in the blaze, told the inquiry in 2022. 'It would have been present. It would have been felt.'[5]

Why was the response to the fire so chaotic? RBKC could have called for external support to help them respond effectively to the blaze, which was undoubtedly too much for one local council to deal with alone. Many other local authorities in London emailed throughout the day offering support with housing or to staff rest centres. There was, in fact, a formal process for activating a pan-London emergency response which could be triggered after serious incidents. But RBKC's chief executive, Nicholas Holgate, chose not to do this. 'That looks like we can't cope,' he said

---

* Renwick D. and Shilliam R., *Squalor*, 2022. The Met Police have denied firearms officers were deployed and said they 'could not account' for the evidence of a witness who said on oath that she saw them.

when his head of emergency planning suggested calling for outside support on the morning of the blaze.

Instead, RBKC smeared the residents. Mr Holgate told a strategic co-ordinating group meeting that there was 'great concern over community tension' with 'hostile residents very vocal in negative comments towards the incident'. A DCLG report on this meeting noted concerns that 'several embittered residents' were 'painting the situation in a very poor light. Incite a mob'. It claimed that 'hurt and anger' was being 'stoked by a small number of known local instigators who continue to fabricate stories to further their aims' and said they may need 'support from the police'.[6] On the ground there was no sign of riots – beyond a noisy protest, mainly attended by outside activists at Kensington Town Hall the weekend after the fire. But central government feared unrest. Mark Sedwill, a senior civil servant and prime minister Theresa May's national security advisor, wrote internally that the fire could become 'our New Orleans' if the response did not improve, a reference to the unrest that followed Hurricane Katrina in 2005.

Amid all the chaos, Tiago Alves, the student who fled the thirteenth floor in the early stages of the fire, had left the flat where his family had been to meet his girlfriend. He realised he'd left his Oyster card upstairs in the tower, which meant it was now destroyed. So he walked a mile or so to the coffee shop where they had agreed to meet, his feet carrying themselves, his mind a blur. 'I was completely on autopilot,' he said. 'I was freezing and boiling at the same time. As soon as I got into Pret I hugged her [his girlfriend] and immediately I just broke down crying.' At around mid-morning, he returned to the flat near the tower where his family had spent the night and saw his sister Ines packing her bag. He asked her where she was going and she said she was off to sit the GCSE exam she had been studying for.

'I said "are you serious?"' he recalls. 'I told her she didn't have to do it, she had extenuating circumstances – but she said she was going to.' She sat the exam and, when it was over, stayed in her seat and burst into tears. Two months later, she would open her results slip to find an A grade.

While the state's response may have been a mess, local community organisations stepped in to the gap. Tiago and several other residents began congregating at the Rugby Portobello youth club on the estate in the days after the blaze. The charity quickly became a space for bereaved and survivors only – excluding the media and the crowds of well-meaning volunteers who descended on the area. It distributed cash to residents to help them replace the possessions they had lost. In the days after the fire, the club cancelled its normal activities and focused wholly on responding to the fire. Among other things, it organised transport to hotels and hospitals, structured clothing provision, a pop-up pharmacy and doctor's practice, provision of smart phones and laptop, and a home pack from Dixons chain store which included items such as TVs, kettles and microwaves.[7]

The Al-Manaar Mosque also opened its doors to survivors. In the hours and days after the fire it opened its doors to survivors, distributing bed linen, clothes, toiletries and other essentials. Volunteers at the centre offered bereavement support through counselling services, and later even acupuncture and art therapy. It also organised interfaith prayers as well as early meetings between the community and the authorities, fire services, police and central government, giving survivors and bereaved answers to some of their questions and an outlet for their anger. Where the state utterly failed, these community organisations provided dignity and support.

The day after the fire, Hassan, the father of Fethia and Hania and husband of Rania, who had been in Egypt caring for a sick relative, returned to London. He was seen by a survivor who knew him in the mosque. The men embraced. 'I was trying to keep it together... I'm the one who broke out crying,' his friend recalls. 'And he told me, he just – his words were, "*Alhamdulillah, Alhamdulillah*" ... "All thanks and praise be to God, whatever we go through." And I think just that kind of reflection of our Islamic tradition, where whatever you go through in life, you deal with it, you have your struggles, you apply the means outwardly, but inwardly... we should strive for that inner peace, where we accept... God's divine decree in things that happen. And for me, I was just – I was blown away just about how he reacted and how he kept it together. That was inspirational.' Recounting this story at a presentation into the circumstances of Rania, Fethia and Hania's death five year's later, the family's lawyer said that while Hassan and his community 'strives for peace as part of their faith' they also continue to 'strive for justice'.

Meanwhile, the condition of some of those in hospital was perilous. Marcio Gomes spent six days being treated. His wife, Andreia, was in a worse condition, due to the smoke she had inhaled on the staircase. She spent fifteen days in a coma. At one stage she had a fifty-fifty chance of survival. Their children were also in hospital, in induced comas. 'We were never contacted by the council, by the TMO. It was just a neighbour who got in touch and asked if we had any clothes or anything. This was after a week,' he said. 'They [the council] said they thought we were staying with family. But they just assumed that.'[8]

RBKC eventually offered to put Marcio into a hotel miles away. He wanted to stay closer to the hospital, but the council said they couldn't help him. 'It was disgusting that they just left me and failed me at my time of need,' he said. In the end the hospital offered him a flat and a charity funded by the McDonald's restaurant chain put him up for three weeks.

During the coma, the doctors told Marcio that Logan, his unborn son, had died. He was stillborn via caesarean section while Andreia remained in a coma. Marcio was there to witness it and held his son's tiny body in his arms. When Andreia came to, fifteen days after the fire, the first thing she asked Marcio was 'How is the baby?' He had to tell her their son had gone. 'It was devastating to have to tell her that he was gone. She had never got to hold him in her arms,' he said.

A NATIONAL SCANDAL

The first major sign that the events at Grenfell Tower were not confined to one tower in west London, but indicative of a much bigger, national problem came nine days after the fire and a few miles away, in north London.

With social landlords frantically checking cladding on their high-rise buildings, blocks were starting to emerge with the same highly combustible ACM as was now known to have been present on Grenfell. Five tower blocks on the Chalcots Estate in Camden were found to have the same material. Not only that, they had also been clad by Rydon and Harley Facades – the team which had led the cladding work at Grenfell Tower. And when the council investigated, they too found problems with fire doors and compartmentation. The blocks were a mirror of Grenfell Tower.

Amid the heightened state of fear which was gripping the country in the aftermath of Grenfell, Camden Council decided it could not tolerate the risk. Overnight on 23 June, residents of more than 700 flats were moved out, with at least a hundred put up in a leisure centre, sleeping on blow-up mattresses provided by the British Red Cross.[1]

It was the start of a national crisis which continues to grip the country today. The government had set up a service to receive samples of cladding from other tower blocks, and every day the number of those affected was growing. A week after the fire, samples from three buildings had failed. With each day that passed, the number grew – five, nine, fifteen. By the next weekend, they had almost thirty.

In the face of these numbers, it appeared obvious that something major had gone wrong. How could our regulations have permitted such a dangerous material to be used on so many high-rise buildings for so long? Reports began to emerge quoting experts saying our rules were more lax than elsewhere in Europe. The Class 0 standard fell under the spotlight, as did the government's apparent failure to act on the lessons of Lakanal House.

But behind the scenes, the government was preparing its defence. It had no intention of simply admitting its regulations were defective and that warnings had been missed. The day after the fire, a public inquiry was announced.[2] While this process would ultimately deliver many of the shocking facts mentioned, it would also ensure any difficult questions would be pushed years down the line. 'We must wait for the outcome of the inquiry,' became the refrain of ministers challenged about apparent government failures.

And the government was also rallying to defend its defective guidance in Approved Document B. Melanie Dawes, the most senior civil servant in the Department for Communities and Local Government (as it was then known), began to learn some of the government's failures as she was briefed after the fire. She said it was the first time she had even heard of the Lakanal House fire – revealing the low priority the blaze had been given within the department in preceding years.

Gavin Barwell, the last minister to be responsible for building regulations, who had personally been given several warnings by the MPs in the All-Party Parliamentary Group campaigning for change, was now chief of staff to prime minister Theresa May. 'Gavin, hope all is ok,' Ms Dawes texted him just hours after the last resident had fled the tower. 'Wanted to let you know that this terrible fire will highlight government delay in changing buildings regs. So you'll want to be briefed personally as ex-housing minister as well as from a Number 10 perspective.'[3]

Brian Martin, the civil servant responsible for building regulations guidance, now found himself at the centre of a storm. This middle-ranking civil servant, whose name was barely known by ministers, knew all about the test in 2001 which had showed the danger of this product, the failure to act after Lakanal and the specific warnings which had been issued and ignored in the years building up to the blaze. But instead of explaining this frankly to his seniors, he told them the material was 'effectively banned'.[4]

This was not how building regulations worked: the law requires builders to ensure the wall 'adequately resists' the spread of flame. It does not prescriptively ban anything. And the presence of the Class 0 standard in the statutory guidance meant the dangerous material was pretty close to being actively endorsed by government. Yet the government would stick to the script that it was banned.

When *The Times* ran a story suggesting otherwise, officials approached independent experts with a pre-prepared script to serve as a rebuttal.[5] At a briefing for the new minister, Alok Sharma, on the Saturday after the fire, a group of experts convened by Mr Martin agreed that guidance banned the material, and advised that its use was likely to be low. The argument was that the word 'filler' in official guidance covered the core of an ACM panel. But Mr Martin knew by now that this meaning was not understood in the wider industry. He had been advised to clarify it and hadn't.[*]

On 22 June, the government made this claim official. In a now notorious letter to housing providers, Melanie Dawes wrote that the words 'filler material' meant 'any element of the cladding system' should be 'limited combustibility'. The government was publicly claiming combustible cladding was banned, when this was patently not the case. At the inquiry, Mr Martin denied this letter was 'a planned, deliberate and underhanded attempt by you and those around you to rewrite history... in order to protect the government's position'.[6] The inquiry will make its own mind up about this in due course. But deliberate or not, the effect was plain. The mistakes of the last thirty years were being covered up.

---

[*] Mr Martin said he was focused on the issues in the immediate aftermath, which is why he did not tell his seniors about previous warnings and issues. He said the statement that all cladding panels were required to be of 'limited combustibility' was included by 'mistake' because he was tired from working long hours during the days after the blaze.

What was harder to cover up was the presence of this material on so many buildings around the country. By 28 June, all 120 samples sent in for testing had failed and the deliveries continued to rise, day after day. The truth was becoming apparent. There had been a desperate failure of building regulations; unsafe cladding was rife across the country and the government would have to act.

The National Housing Federation – the body which represents housing associations – called for 'immediate, decisive and strategic action to ensure the safety of buildings across the country'. 'This is one of these occasions of such magnitude that only government can take this responsibility on behalf of us all. This will cost a lot of money. The nation can, and must, afford it,' it's chief executive wrote.[7] But this is not what was planned.

On 27 June, the government had appointed an expert panel to advise it on the correct response. This would be chaired by their long-time advisor Sir Ken Knight and feature Dr Peter Bonfield, the chief executive of the BRE.[8] They recommended a series of tests on systems containing ACM to be carried out at the BRE's labs. These were carried out over the summer, with four systems experiencing serious failures. A system similar to the one used on Grenfell Tower failed dramatically, with flames reaching the top of the 9m rig after just seven minutes. But the expert panel did not order their removal. And the government did not announce a penny in funding to fix blocks with these panels. Instead, the official advice to building owners was to 'take their own professional advice'.[9] The government did not see this as its problem to solve.

It also took steps to stop the full scale of the crisis becoming clear. While regular updates were given about the number of affected buildings, it was not possible to discover where they were. The government, under the pretext of national security, claimed that terrorists might set light to the cladding if details were released. As a result, it ordered local authorities not to release details about the buildings under the Freedom of Information Act. Whether or not the intention of this step was genuine, it meant the media coverage in the months after the fire was muted, the outrage absent. It is likely that many of those who lived in the buildings never knew the full truth about why scaffolding was being rapidly erected outside their windows.

The government also seriously underestimated the scale of what was about to emerge. Up until this point, the focus had been on social housing. Somehow, the government had convinced itself that the problems would be limited to housing for the poorest. 'The question I keep coming back to is very simple,' said secretary of state Sajid Javid in September 2017. 'Would a fire like this have happened in a privately owned block of luxury flats?'

Suggesting the answer was no, he said the fire signalled the need for 'a fundamental rethink of social housing in this country'. But Mr Javid's confidence in the private sector was misplaced. The effect of deregulation had been so deep and the construction sector so out of control, that consequences spread well beyond a few hundred social housing blocks. This crisis was about to emerge. And the failure to deal with it was to bring misery to hundreds of thousands of flat dwellers all across the country.

Ritu Saha and her husband had come to England from India in 2009. 'We wanted a bit of an adventure,' she says. 'We wanted to travel all over Europe.'

The couple rented in Ealing for a few years, putting money to one side to save for a deposit on a flat. They applied for permanent residency and then citizenship. 'Because I come from quite a conservative financial mindset, I wanted to be debt-free as soon as I possibly could,' she said. 'This was the first debt I had ever taken.' The couple found a flat in a block, Northpoint in Bromley. 'We completely fell in love with the flat,' she recalls. They got the keys in December 2015. 'That was an extremely happy time for us. Some little girls dream about having a family, I used to dream about having my own house,' says Ritu. 'I had saved a bit of money to buy nice furniture. I knew exactly what I wanted my house to look like.'

In June 2017, along with the rest of the country, Ritu saw the images of the fire at Grenfell Tower. 'I did not think for a minute that what happened at Grenfell could happen in my building,' she says. 'I had lived there for one and a half years, the building was very well maintained.'

But in November 2017, a letter landed from their managing agent saying they were investigating their cladding, and a patrol of wardens was being put in place at the request of the fire brigade. The letter said they would be gone by January. But just after the New Year, a neighbour knocked at her door and asked her if she knew the bored-looking security guards sitting at reception were costing them £6,500 a week collectively in service charges. 'That was when I first realised there was something big happening,' she says.[10]

In the UK, private flat owners don't really own their flats. Instead, they buy a long lease on the property, and the building remains owned – usually – by a private company. These landlords might have little interest in building management, and hold the property mostly as a useful source of income for their investors. But despite not owning the flat, English law makes leaseholders financially liable for the upkeep of the building. And as it began to emerge that cladding would need ripping off the walls of hundreds of private blocks, this created a problem. The bills for the work were vast. They often came to around £30,000 each, but in the most extreme cases stretched to six figures. This was money people simply did not have. And on top of that, they were also suddenly asked to pay an extra £500 a month for the cost of permanent security guards, known as 'waking watches', to evacuate the building in a fire. And as their insurers woke up to the risk of the combustible cladding on their walls, the premiums soared – rising by more than 1,000% in the most extreme cases.[11]

At first residents tried to go to tribunals, certain that such unfair bills could not legally be placed at their doors. But one after another they lost. The law may have been unfair. But it was the law. By March 2018, more than three hundred buildings had been confirmed as having a dangerous ACM system. Only three had been fixed. More than a hundred were private buildings where leaseholders were on the hook. 'This will become a huge national crisis unless the government acts quickly,' said Lucy Powell, the Labour MP for Manchester Central.[12] They did not.

Under pressure from survivors and bereaved, who had formed lobbying group Grenfell United featuring Eddie Daffarn, Natasha Elcock and Tiago Alves as key members, Theresa May finally committed to stumping up some money in May 2018. She offered £400m to pay for the remediation of social housing blocks – cash she top-sliced from the government's pot of money to build new afford-able housing. But she did nothing for private blocks. The government would not step in to pay for the repairs of privately owned homes. Still in denial about their own responsibility for Grenfell, they insisted it was not their fault. The best they could offer leaseholders was a weak-sounding and often repeated plea to private building owners to 'do the right thing'. They insisted that – thanks to the waking watches – the buildings were safe to occupy.

At Northpoint, as the waking watch remained in place, the cost pressure simply became too much for residents to bear. Ritu's neighbours decided they would keep watch themselves. She spent Christmas of 2018 wearing a high-vis jacket, walking the empty corridors of her flat, checking for any signs of fire. And Ritu was not alone. The number of people affected was about to grow exponentially.

While ACM is probably the most dangerous of the cladding materials to have been fitted to the walls of our homes in the last thirty years, it is not the only one. There is combustible insulation, with its potential to produce choking, toxic smoke and various other products: rendered polystyrene, timber cladding and balconies and high-pressure laminate (HPL, thin wooden boards laminated together with glue – the material used on Lakanal House). There were many other buildings with missing cavity barriers – a product of the same poor quality workmanship evidenced on Grenfell Tower. In autumn 2018, the government saw testing which showed the catastrophic failure of a cladding system made of HPL panels and combustible insulation. This is a combination which is widely-used in the real world. They doubled down on their claims that anything combustible was banned by guidance. They said anything that was combustible and had not passed a large-scale test needed to come off buildings.

The effect was chaos. Now every building in the UK needed to establish whether or not it had combustible materials in its facade. There were not the assessors nor the testing facilities to do this. But by now, lenders had woken up to the problem. They did not want to provide mortgages on flats that would require expensive repair work or – worse – be destroyed in a fire. They began to refuse mortgages unless compliance could be shown with government guidance. The flat sale market ground to a halt.

People were told that their flats were worthless until expensive remediation was carried out. This meant they could not sell. Amid a series of miserable stories about people who could not move to be with sick relatives, or upscale to have children, one resident sobbed as she told the BBC that her home was now 'a prison'. 'I just want out,' she whispered.[13] The mental health toll on those living in affected buildings was severe. A survey published in spring 2019 revealed 64.8% of respondents said their mental health had been 'hugely affected', 69.5% said they feel anxious or worried about the future on a daily basis, 81.6% had experienced stress, 77.6% anxiety, 65.8% difficulty sleeping, and 23% had fallen into depression and 8.7% were experiencing suicidal thoughts.[14] 'The effects of the uncertainty are unimaginable unless you're in the situation yourself. I wish more than anything I hadn't worked hard to save for and buy a property ever,' said one respondent. 'I feel constantly stressed, anxious, depressed, lost, abandoned and devastated.'

However, Ritu was not someone who would simply accept this situation. She began to talk to the media, appearing in the *Guardian, Inside Housing* and on ITV. One *Guardian* journalist she spoke to had an idea – why not form a campaign group to take up what was clearly a national issue?

She took up the idea mostly because it gave her a better chance of prominent coverage in the paper. Together with a handful of other residents, who had connected over Twitter, she announced the formation of the UK Cladding Action Group.

In Manchester, a similar process was underway – with residents of affected blocks coming together and beginning to realise that they had a fight on their hands and it was one that needed to be fought together. They formed a group called Manchester Cladiators. Together, the two organisations then approached *Inside Housing*, which had published several stories about the issue and we supported them in a joint launch of a new campaign 'End Our Cladding Scandal' in April 2019. With the backing of the bereaved and survivors at Grenfell Tower, the campaign was able to push the government to release £200m. This would cover the ACM removal for high-rise buildings. But everyone else would be left to fend for themselves.

In the months that followed a series of fires spelled out the danger of this approach. On 9 June 2019, as the second anniversary of Grenfell approached, a block of flats with highly combustible timber cladding was utterly consumed by fire when a barbeque on a balcony tipped over. Everyone escaped – but it was a close call for many. Some had to kick down magnetic doors which would not open. One man climbed down from his fourth-floor balcony with his pet cat in a box. The fire gutted the building. Residents, many of whom had no contents insurance, lost everything. Many pets were burned alive.[15]

In November, fire ripped through another block – this time The Cube student accommodation in Bolton. This block was clad with high-pressure laminate panels. After starting due to a flicked cigarette on a balcony, fire tore through the facade of the building. Greater Manchester Fire and Rescue Service was able to evacuate the block but it was, yet again, a close call.[16]

And against the backdrop of these fires were tens of thousands of furious, distraught leaseholders – unable to sell, seeing their savings bled dry and facing the crushing future possibility of a bill that would bankrupt them. A government-backed effort by the industry to fix this

– EWS1 forms which could be signed by a surveyor to confirm the safety of the block for a lender – backfired. Offered the chance of security, the mortgage industry began demanding them for blocks of all heights – even occasionally on converted terrace houses. Thousands and thousands more blocks joined the queue for remediation.

In the years that have followed, under immense and sustained pressure from the resident campaigners, there have been some slow improvements to this position. The government has steadily announced more and more money to cover the remediation of dangerous buildings, with the current sum reaching £9bn – including £4bn which it hopes to raise from industry contributions. And despite voting down proposals several times, it finally passed a law in early 2022 which gives residents of many buildings at least some protection from the costs of fixing them – with new processes set up to recoup the money from other responsible parties. But this process has been fraught, slow and miserable. Many lives have been irredeemably ruined by the stress – I have heard stories of marriages which have broken down, life savings lost, bankruptcies, careers ruined and the chance to have children lost. There are even reports of suicides of some of those for whom the pressure of living in a dangerous building, unable to afford the cost of fixing it, simply became too much.[17] And five years on from the Grenfell Tower fire, for all this pain, many buildings remain unfixed. While most of the ACM cladding on high-rises has been removed, or at least seen work begin, just thirty blocks out of more than three thousand that registered for government funding due to other issues have completed. For medium rise buildings, we still do not even have clear figures. The government's best estimate is that between 6,620 and 8,890 buildings will require work to fix them due to 'life safety' risks, likely accommodating up to half a million people. Many have not even been identified, let alone had work on them begin.[18]

And cladding removal is, sadly, just one of the ways in which we have failed to respond to the Grenfell Tower fire.

*

If ripping combustible materials off the walls of buildings is a complex, expensive challenge, surely not sticking these materials to the outside of new ones is much easier. But after Grenfell, even banning the use of combustible materials on new buildings became complex.

Kingspan, the insulation manufacturer, set out to lobby against such a change to the rules just six weeks after the fire. It hired a large public relations firm to help it lobby to ensure any changes did not damage its business interests. The firms drew up a list of influential MPs to target with its message: that combustible insulation could be used safely on high-rises so long as it was properly installed.[19]

Leaked minutes from a high-level meeting of the firm in September 2017 spell out this strategy. 'Support [large-scale] tests, support desktop studies,' the notes said. It also said the firm would be 'challenging the norms that are associated with non-combustible'.[20]

The government, meanwhile, hired an independent expert Dame Judith Hackitt, to review building regulations and make recommendations for change. But her review, published in spring 2018, was remarkably kind to the government. While she said there had been 'a race to the bottom' in the construction industry, she adopted without question the official position that combustible cladding was effectively already banned by the official guidance and that the only reason it was on buildings was non-compliance.

As a result she did not recommend substantive changes to the rules, and instead said new processes should be introduced to more effectively share information between the teams involved in a construction job, and determine more clearly at each stage who was responsible for safety. These changes are welcome, but without tightening the rules, they are no guarantee of safety. Her review recommended, effectively, sticking with the status quo produced by deregulation. 'The aim of this review is to move away from telling those responsible [for tower blocks] "what to do" and place them in a position of making intelligent decisions about the layers of protection required to make their particular building safe,' she wrote.[21] As such, she rejected the need for an overt ban on combustible materials.

This was immediately pilloried by Grenfell United. The group had met with Dame Judith before the publication of the report. Their knowledge of the gaps in the relevant regulations was advanced and they were unequivocal: they wanted this material banned. 'We know dangerous cladding costs lives,' said Natasha Elcock who was by now balancing her supermarket job with leading the group. 'That's why the Hackitt report should ban it from ever being used again.'[22]

Under pressure from the group, the government U-turned. It announced plans to ban the use of combustible materials on buildings above 18m in height. This would be the first piece of prescriptive building regulation introduced since Thatcher.

But it only went so far. Buildings below 18m were still effectively subject to a free-for-all – even the limited pre-Grenfell standards in guidance did not apply to these buildings, which could be up to six storeys high. This led to the bizarre position where the risk averse stance taken by lenders meant medium rise blocks were being subject to expensive demands to have combustible materials ripped off, while at the same time 75% of new buildings below 18m contained combustible products in their facades.[23] It took until May 2022 for the government to change the regulations in this area. And even then, it was not willing to introduce a total ban. It said that blocks between 11m and 18m could only use combustible materials in a system which had passed a large-scale test.

But Grenfell was never just about cladding. As the fires in other parts of the world prove, cladding fires do not have to involve mass fatalities if people can get out of the building. As well as calling us to strip dangerous cladding from the walls of high-rises, Grenfell also demands that we step back from our total reliance on stay put. In this area too, we are failing.

After hearing the evidence of the night of the fire, the Grenfell Tower Inquiry published its first phase report in October 2019. The key change the inquiry called for was for building owners to move away from their absolute conviction in 'stay put'. Instead, the report said, building owners should develop a Plan B. Evacuation plans should be developed for all buildings in case compartmentation were to fail unexpectedly. They should also be fitted with manual alarms which would allow the fire services to send an evacuation signal to all or part of the building, if required. And disabled residents should be offered personal emergency evacuation plans (PEEPs).[24]

The government immediately committed to implementing the report in full. But behind the scenes, there was pushback. In April 2020, a private meeting was held with representatives of the fire and housing sector and the Home Office to discuss the implementation of the inquiry's proposals – to which the government had publicly committed. Minutes (which I later obtained) record that the group said the provision of PEEPs in all high-rises would be 'completely impracticable and not doable'. 'Stay put is conducive to people who are vulnerable – it is safe for them to stay in their flat,' the minutes say – two years after 41% of the disabled residents of Grenfell Tower died after being told to stay put. General evacuation plans were also dismissed. 'It is difficult to see what an evacuation plan might comprise, other than a simple reiteration of the standard stay put strategy,' the minutes said. Alarms were also questioned on the basis of 'cost against benefit (it's not cheap)'.[25]

The government was evidently listening. In summer 2020, its consultation rowed right back on the idea of PEEPs, suggesting they should be used only in blocks known to have dangerous cladding. This infuriated the bereaved, with the children of Sakina Afrasehabi launching a judicial review to force the government to honour its promise to provide PEEPs to all residents. Faced with an embarrassing legal action, the government backed down and promised to reopen the consultation.

The consultation produced overwhelming support for PEEPs, with 83% of respondents agreeing that they should be introduced. But the government did the opposite. After its consultation closed, it held one-to-one sessions with building owners who raised concerns that the process might involve the imposition of full-time staff and therefore prove too expensive. As a result, in May 2022, the government announced it would not follow the inquiry's recommendation. Claiming it would not be 'proportionate' to require building owners to prepare evacuation plans the government even said such plans wouldn't be 'safe', because disabled residents could slow the escape of able-bodied people. Instead, the government wanted disabled residents to continue to rely on stay put in most buildings. Those living in buildings with dangerous cladding would have their details shared with local fire services and no further plans would be made for getting them out of the building. Bluntly, policy dictated that if they were trapped by a fire before the fire services got to them, they would die.

Sarah Rennie, a wheelchair user who lives in a block of flats with dangerous cladding and campaigns for evacuation plans for disabled residents, said:

'I've always been disabled, and so ever since I was in school, I have experienced the fire alarm blaring and going to sit in a stairwell while everyone around me leaves for safety. As a child and a young person, I accepted it. I thought that was my lot. Like disabled people across the country, I sleep feeling anxious, frightened, and today frankly traumatised to read what the government's written today that it believes everybody's lives are more valuable, because I could get in the way of them getting out.'[26]

This attachment to stay put – on which we are a global outlier – remains so strong that we persist in not implementing other measures of keeping people safe in a fire. We continue, as said, to be one of the only countries in the world, along with South Korea, that allows buildings to be constructed at any height with a single staircase. We continue not to have fire alarms in almost all residential blocks. And there remains no co-ordinated effort to retrofit sprinklers into high-rise buildings, despite their recommendation from the coroner investigating Lakanal House back in 2013.

Instead, we seemingly believe that all that was wrong with stay put was cladding. And so, we have pursued a botched attempt at removing it from all the UK's buildings instead of investing more broadly in introducing more layers of protection and giving people the means to evacuate quickly if a fire gets out of control.

This means we cannot yet say that another Grenfell is impossible. Cladding was what experts described as a 'blind spot' in our plans for fire safety in the years before Grenfell. It shouldn't have been, but it was. We believed our buildings were safe and missed this critical risk. Even if we do manage to get the most dangerous cladding systems off other buildings, there will be other blind spots. Timber framed buildings can carry terrifying dangers − where fire can break into the space behind walls and ignite the entire structure of the block. We have seen large buildings burned utterly to the ground in fires like this, several of them since Grenfell. The same is true of the increasingly in vogue 'modern methods of construction'. These are blocks of flats constructed of dozens of factory-built units, which are then quickly assembled on site. But if gaps are left in the internal structure and not properly protected, fire could rip through the inner body of the building, putting the entire structure at risk. Such a blaze happened at a hotel in the Shetland Islands in 2020 − leaving the entire building a heap of burnt rubble.[27]

What we need is the same as what we've always needed. Robust regulations and enforcement to prevent bad practice, but a Plan B and Plan C in place for what to do if these protective measures fail.

The world is getting hotter. Hot summer days provide a perfect tinderbox for fires. There will be more to come. And despite everything we have seen and all the suffering at Grenfell Tower, it remains entirely possible that we will once more witness the terrifying sight of a building being consumed by fire with people trapped inside. Government officials are alleged to have said 'show me the bodies'. They have seen them now, and still we have not changed.

'KEEP GRENFELL IN YOUR HEARTS'

As the story of Grenfell Tower has morphed into a national political scandal about who will pay to fix other tower blocks, it sometimes feels as if the community to whom this happened have been forgotten.

But their five years has also been one of suffering. As the immediate storm of media interest in the blaze faded, the community began to face up to their loss. The survivors and bereaved were not only struggling with grief and anger, but also the ongoing failure of the rehousing process which meant many of them were still stuck in hotels or temporary accommodation. Many were diagnosed with post-traumatic stress disorder or other mental health conditions linked to fire.[1]

There were also mounting concerns about physical health. A condition known locally as the 'Grenfell cough' was afflicting many of those who had breathed in the choking smoke inside the tower, and many others who had watched from the ground or continued to live nearby.[2] For these residents, the daily struggle of living in the shadow of the tower where they had seen so many friends die was enormous. 'You feel like it's going to fall on you,' one of these residents told *Inside Housing*. 'I feel like crying every day. Every day in this place, my skin is up. The children tell me, "Mama, I saw a ghost." '[3]

Meanwhile, funerals were being held for those lost in the fire. Debbie Lamprell's friends from her childhood in north London flocked back to the capital to say their goodbyes. Singers and orchestra players from the Holland Park Opera, where she had worked, turned out. When the procession passed the school where her mother had worked as a dinner lady for a generation, the whole school came out. Her ashes were laid in the City of London Cemetery and Crematorium next to her beloved father.

'I don't really know what made her so positive. It's not that she had so much money or anything; but she had her freedom, she did what she wanted to do and she loved people. And I think that made her rich,' her mother reflected later in a statement to the inquiry.

'I am an old woman with nothing else left. And maybe it's taken losing Debbie to realise we weren't normal. Debbie was an exceptional, extraordinary person, and I was completely blessed to have her as my daughter.'

Rania's sister Rasha flew to London. 'I was still not convinced that she passed away, I was searching for her in the faces who passed my way, until I saw her remains in the morgue and I wish I never saw her there,' she recalls.

'I asked to stay alone with her and I heard her tell me she was well and that I will continue in her path. I felt it was real and not my imagination and her promise that I will finish her way.'

Years later, she told me she still believes her sister is with her. 'I still speak to her when I go through difficulties and her spirit always assures me that all difficulties will end. I am still going ahead, I learnt how to swim and ride a bicycle and lost some weight like she was doing. She still is with me, and still my mind goes numb when someone talks about her as a dead person, I feel Rania is in my heart but no one knows except me.'

When Rania was young, she had a huge crush on Prince William. She used to tell Rasha that she would marry him when she grew up. When Rasha saw the prince attending a memorial ceremony for those lost at Grenfell Tower, she laughed and cried. 'Who is this girl that whenever she wished for something, she got it in a way or another,' she said. 'That is Rania.'

*

As weeks and months passed and the first anniversary of the fire approached, it was clear that efforts to rehouse those who had fled Grenfell Tower were failing.

In a nutshell the problem was that while many properties were bought for the people of Grenfell Tower, they were not bought *with* the people of Grenfell Tower who were expected to effectively take what they were given. Theresa May boasted in the week after the fire that the government had secured 164 properties for survivors. But many proved inappropriate. For example, homes located in the richer part of the borough would have meant residents on lower incomes could not afford childcare and could not get their children to school. Or properties did not have the right adaptations for disabled residents. Or they were in high-rise blocks, which traumatised children refused to set foot in. One of the blocks of homes offered to the community was found, after twenty survivors had moved in, to have 'high' fire safety risks – including defective fire doors and patches of suspect cladding on the outside walls.[4]

A bidding process pitted survivors against one another for the best properties. Other families were offered homes with damp, or strange smells. They declined private rented accommodation, fearing they would lose their secure council tenancies if they accepted. The net result was traumatised families spending months making do in hotel rooms, sofa-surfing or in temporary accommodation.[5] This was often difficult. Bellal El-Guenuni, for example, was sent to a hotel room with his wife after escaping the tower. But when their children were discharged from hospital, the hotel manager tried to throw the family out, claiming there were too many people for the room. By the first anniversary, sixty households from the tower remained homeless.

Nick Burton meanwhile was supporting his wife Pily. In the confusion after the fire he had been taken to a different hospital to his wife, but he had eventually found her in the Royal Free Hospital in north London. Her condition was serious – like most of those who escaped the tower late into the night she had suffered serious smoke inhalation on her way down the stairs. In an already fragile state, her recovery was difficult and the trauma of the fire had impacted her dementia. 'How do you explain what had happened to a person in her condition?' Nick said later at the inquest. 'That her house had gone, her dog had gone, her good friends and neighbours may have passed, and many friends were missing. And that her parents' ashes, which we kept in the flat, had gone. Everything was gone. It was just too much.'

In January 2018, she suffered a stroke. After two weeks on life support, doctors gently told Nick there was nothing they could do for her. He moved a mattress into her hospital room and stayed there, day and night, holding her hand for a week. On 29 January, just as her son walked into the room, she drew her last breath. Nick said it was like she had been waiting for him.

'All that life, the essence of a person, how they lived their life, was gone. It doesn't matter what religion or belief you have, it's just a very sad moment,' he said. 'We don't know what life is, really. We don't really know what we are or who we are or what life has in store. But we're just truly blessed, and I'm truly blessed that I had thirty-four years with such a person. We had fun. We laughed. We got to go through life together and it was wonderful. And this is all we can do.'

*

The community has also had to grapple with the inquiry, and the seemingly endless legal treadmill on which they are now trapped. The process began with a series of commemorations to those lost in the blaze. Families were invited to give presentations on their lives. They were funny, touching, heart-warming, heart-breaking and angry. None have stayed with me quite as much as the words of a father describing the loss of his five-year-old son. He said his son would have lived if they had not been told – by a firefighter who reached them at 2 a.m. – to stay put in the flat.

'Every single minute from that day, I'm thinking: what if we had not waited? What if that fireman at around 2 a.m. has taken us out instead of telling us to wait and shut the door? What if? And this "what if" makes my head and heart explode with pain and anger, and my heart is shredded. But I have to keep smiling, don't cry, don't show your emotion, be strong, be a man, be a strong father. My life is in ruins. We cannot even bear to talk about what happened. [My younger son] looks away if he sees [his brother's] picture. I cannot bear to look at myself in the mirror. His mother has her church and her faith to keep her going on, and I was only able to say my beautiful son's name just three weeks ago.'

Over the years, they have listened as the corporations and public bodies responsible have given carefully constructed statements, full of vapid expressions of sympathy and detailed explanations of why their company is not to blame for what happened at the tower. The inquiry's own barrister has described the attitudes as 'a merry-go-round of buck passing'.

They have also had to see the revelations pushed out of the public eye, as Brexit negotiations, political turmoil, Covid-19 and war distract attention from the evidence that has slowly played out in the hearing rooms in west London.

When Eric Pickles, the minister to whom the Lakanal House coroner's letter was addressed, finally gave evidence in April 2022, he berated the barrister questioning him for spending too long on a topic saying he had 'promised we will be away this morning' as he had 'an extremely busy day meeting people'. He then closed his evidence by paying tribute to the 'nameless 96 victims' of the fire. He would later admit he had mistaken the tragedy with the Hillsborough stadium crush in 1989.*

What many are desperate to see is justice. But that remains some way off. An active police investigation has interviewed more than forty people, some of them under caution, but arrests will not happen until the inquiry report is published – which is unlikely to be until well into 2023.** Even then, trials will take years and other atrocities like Hillsborough prove there is no guarantee of success. The road is long and uncertain, and all the while they have to carry the burden of their grief, their anger and their terrible loss.

I spoke to Marcio Gomes in the summer of 2021. His family had a new home, but the weight of what had

---

* Inquiry transcript, 7 April 2022. The correct death toll at Hillsborough is 97, following the death of Andrew Devine in 2021. Lord Pickles apologised for being 'discourteous' during his evidence and for offending the bereaved by getting the number of victims wrong.

** In February 2020, the UK's Attorney General agreed to protection against self-incrimination for witnesses at the inquiry. This means oral answers cannot be used by prosecutors to further advance a case against the individual who gave them. This was widely and inaccurately reported as immunity from prosecution. However, prosecutors can use all the other evidence to bring charges.

happened still stayed with him. He recalled playing football and suddenly finding himself completely out of breath.

'I saw a doctor and they explained that it was my mind,' he said. 'When I started to feel exertion, it brought back the memories of escaping the tower. And because my body could remember all the smoke I had inhaled, it shut down and limited my breathing.'

He had to see a specialist to help him 'learn to breathe again'. He said the place that the family had been rehoused was nice, but it wasn't home. 'Our home was the tower,' he said. 'And that's gone.'

As for Nick Burton, he went on a round-the-world trip after Pily died. In each country he visited, he went to the fire station and spoke to them about Grenfell. 'Everywhere I went, they had heard about it and they wanted to invite me in and talk,' he says. His pilgrimages were marked by a desire to make sure other fire authorities would not make the same mistakes. He would ask them what their plan was for a burning building and what sort of cladding there was on local high-rises.

Back in London, he had a nice ground floor flat with a small garden, around a mile from the tower. It had neat decking in the back garden and vibrant pieces of modern art on the walls. But the first thing Nick showed me was a broken, burnt out laptop. It had recently been released by the police, and was the largest item he was able to recover from the flat in which he had spent so many happy years with Pily. 'I haven't got used to being on my own,' he said.

On the fifth anniversary of the fire, the community held a silent walk around the streets near the tower, followed by a rally at its base. Speakers called for justice, for arrests, for change. Karim Mussilhy, who lost his uncle in the blaze said:

'It's been five years, and we're still here saying the same things. I can't help but think about the kids that we lost. When I close my eyes, I can still hear them playing. And we're no closer to justice than we ever was in the first days.

But what have we learned? It's quite simple. The government do not care about you. They never have and they never will. It's that simple… You've got to see yourself in us. You've got to realise that what happened here wasn't some freak accident. It was an easily preventable thing they allowed to happen. Keep Grenfell in your hearts. The issues are still here today. Make sure this never goes away.'

*

There are many points in history at which the story of the years after Grenfell could end. But the one I will choose is 6 p.m., 22 June 2020, on the twelfth floor of a high-rise block in Canning Town, east London – only a stone's throw from the site of Ronan Point, which suffered the collapse in 1968.

Like Grenfell Tower, Ferrier Point had been clad in ACM during a refurbishment project. The work was done by Rydon and Harley Facades – the same duo which refurbished Grenfell Tower. Like Grenfell, Ferrier Point had serious internal fire safety defects – deemed 'appalling' by experts including Arnold Tarling in a report for the BBC in the days after Grenfell.[6]

On 22 June 2020, just over three years after the Grenfell Tower fire, a blaze started on the twelfth floor of Ferrier Point. It smashed a window, and smoke billowed up into the air above east London, flames lapping against the external cladding.

But that is where the fire stopped. Thanks to the lobbying of the survivors and bereaved of Grenfell Tower, the government had reluctantly put up £400m to remove ACM cladding from other high-rise blocks in May 2018. Ferrier Point was one of the blocks to benefit from this funding. The work to remove and replace the cladding had been completed only months previously. More than 150 residents evacuated the building, but none were hurt. The fire was under control by 7.45 p.m. and most of them returned to undamaged flats.[7]

Here is one small positive lesson we should take from the aftermath of Grenfell. Amid seemingly endless frustrations and knock-backs, the survivors, bereaved and campaigners for justice have had some victories.

The world that gave us the Grenfell Tower fire looks irredeemably dishonest. It is a story of corporate structures that allowed human beings to abandon their own conscience and sense of agency and to think only about sales and profit margins. Government institutions placed ideology above human lives at every turn. Listening to the evidence of the people within these structures over the years, one leaves with a profound sense that they felt trapped. They simply served as drones, turning the wheels of a machine far beyond their control.

But there is another vision of humanity available from the Grenfell Tower tragedy. It is of those who stopped as they fled the burning building to help their neighbours flee. Or of the firefighters that went back into the fire, up the smoke-filled, dangerously hot staircases, with no water, faulty radios and no guarantees that the building was not about to collapse around them. It is the mosques, churches and voluntary organisations that provided some semblance of a humanitarian response as the state disgracefully failed to respond. It is those who stood up on behalf of their neighbours and demanded better treatment from their landlord before the fire and continue to fight for others after it.

As we stand on the brink of a century when the consequences of the same deregulated economy that gave us Grenfell threaten to set the whole world on fire, it is a fight we must all join in some way. And in any small victories that follow, we can at least glimpse the shadow of a better world.

22

THE DECEASED

Many names are missing from the account of the fire you have just read. I have only included in the preceding chapters the accounts of victims for whom a next of kin has given express permission. Understandably, some did not feel comfortable. Others were not easy to track down and it felt better to leave them to their privacy and their grief. But it would be wrong to write a story of Grenfell that does not remember every victim lost that night.

There were seventy-two. Eighteen of them were children. Each was loved. Each is missed, desperately, by those left behind. Each suffered unspeakably for the mistakes, greed and callousness of a system that turned their home into a death trap. These are their names:

Fathia Ahmed Elsanousi, Abufras Mohamed Ibrahim, Isra Ibrahim, Mohammed Amied (Saber) Neda, Hesham Rahman, Rania Ibrahim, Fethia Hassan, Hania Hassan, Marco Gottardi, Gloria Trevisan, Raymond Herbert (Moses) Bernard, Eslah Elgwahry, Mariem Elgwahry, Anthony Keith Disson, Bassem Choukair, Nadia Choucair, Mierna Choucair, Fatima Choucair, Zainab Choucair, Hashim Kedir, Nura Jemal, Yahya Hashim, Firdaws

Hashim, Yaqub Hashim, Sirria Choucair, Abdulaziz El Wahabi, Faouzia El Wahabi, Yasin El Wahabi, Nur Huda El Wahabi, Mehdi El Wahabi, Ligaya Moore, Jessica Urbano Ramirez, Omar Belkadi, Farah Hamdan, Malak Belkadi, Leena Belkadi, Mary Ajayi Augusta Mendy, Khadija Saye, Victoria King, Alexandra Atala, Mohamednur Tuccu, Amal Ahmedin, Amaya Tuccu Ahmedin, Amna Mahmud Idris, Majorie Vital, Ernie Vital, Debbie Lamprell, Gary Maunders, Berkti Haftom, Biruk Haftom, Hamid Kani, Isaac Paulos, Sakina Afrasehabi, Fatemeh Afrasiabi, Vincent Chiejina, Khadija Khalloufi, Kamru Miah, Rabeya Begum, Mohammed Hamid, Mohammed Hanif, Husna Begum, Joseph Daniels, Sheila, Steven (Steve) Power, Zainab Deen, Jeremiah Deen, Mohammad Alhajali, Denis Anthony Peter Murphy, Ali Yawar Jafari, Abdeslam Sebbar, Logan Gomes, Pily Burton.[1]

ACKNOWLEDGEMENTS

There are many people who have contributed to this book. First – thank you to the team at Oneworld for their unwavering belief in the project and hard work on it. Thanks particularly to Alex Christofi for seeing the potential in it. Thanks to Cecilia Stein for good advice at many stages and sympathy and understanding when it proved difficult. Thanks to Rida Vaquas, Paul Nash and Holly Knox for their forensic and skilled work on the manuscript. Thanks to Dan and Ines for making an important introduction.

Thank you to everyone at *Inside Housing*. Our reporting on fire safety and the Grenfell Tower Inquiry has always been a team effort and this book would not be possible without all of your efforts. Particular thanks to Martin Hilditch for wise counsel, faith in me and willingness to let us pursue this story wherever it led. Thanks to Jack Simpson, Luke Barratt, Sophie Barnes, Nat Barker, Grainne Cuffe, Lucie Heath and everyone else who helped the mammoth task of reporting the inquiry and so much else besides.

Thank you to the BSR and community lawyers whose work on the inquiry has been extraordinary and facilitated the release of many of the details which are relied on here.

Thanks to Phil Murphy and Remy Mohamed for reading early drafts and providing extremely useful feedback. Thanks to Jonathan Evans, Ian Abley and Andrew Chapman without whom I would never have understood the intricacies of Approved Document B. Thanks to Arnold Tarling, Sam Webb and the rest of your gang, who have been fighting the battle for safer high-rises since before I was born and without whom *Inside Housing* would have been much slower to address these issues.

Thanks to Deborah for important and professional support at a key moment.

Thanks to Lucy for all her belief in me, advice and support. This would never have happened without you. Thanks to Mum and Dad for everything. Thanks to Jalia for your patience and love. Thanks to Benjamin and Samuel for tolerating Daddy working, instead of playing, far too frequently.

Finally, thanks to all those in the Grenfell community who have trusted me to tell this story – especially those who spoke to me directly about their experiences or granted permission to tell the story of their loved ones. I sincerely hope my book repays that faith and I hope the justice you are all waiting and hoping for comes soon.

NOTES

Introduction

1 Davenport, J. et al, '40 lives saved in Camberwell Tower inferno', *Evening Standard*, 6 July 2009. https://www.standard.co.uk/hp/front/40-lives-saved-in-camberwell-tower-inferno-6784010.html

2 Inquiry transcript, 6 December 2021.

3 Barnes, S., 'Government delay in reviewing fire safety regulations "putting tower blocks at risk"', *Inside Housing*, 7 March 2017. https://www.insidehousing.co.uk/news/news/government-delay-in-reviewing-fire-safety-regulations-putting-tower-blocks-at-risk-50024

4 Apps, P., 'A stark warning: the Shepherd's Bush tower block fire', *Inside Housing*, 11 May 2017. https://www.insidehousing.co.uk/insight/insight/a-stark-warning-the-shepherds-bush-tower-block-fire-50566

Chapter 1

1 Conversation abridged.

2 *Grenfell Tower Inquiry Phase 1 report*, volume four, page 515. The report is publicly available here: https://www.grenfelltowerinquiry.org.uk/phase-1-report

3 Ibid., page 515.

4 Behailu Kebede, inquiry witness statement. https://assets.grenfelltowerinquiry.org.uk/documents/Inquiry%20Witness%20Statement%20of%20Behailu%20Kebede%20and%20accompanying%20Exhibits%20BK%3A1%20-%20BK%3A3%20IWS00000490_0.pdf

5 Slater, B., 'Revealed: the brands linked to the most appliance fires', *Which?*, 15 February 2018. *https://www.which.co.uk/ news/2018/02/revealed-the-brands-linked-to-the-most-appliance-fires/*

6 Dr John Glover *Phase 1 report*. https://assets.grenfelltowerinquiry. org.uk/documents/Dr%20John%20Duncan%20Glover%20 report%20%28Phase%201%20-%20supplemental%29%20 JDGR0000001.pdf

7 Expert evidence, 27 November 2018, page 133, line 20. https:// assets.grenfelltowerinquiry.org.uk/documents/transcript/Expert-Evidence-27-November-2018.pdf

8 https://www.judiciary.uk/wp-content/uploads/2014/07/ Lapping-2014-0214.pdf

9 Where uncited, quotes are taken from the witness statements of firefighters or residents given to the first phase of the inquiry.

10 *Phase 1 report*, volume four, page 593.

11 Howkins, R., 'Concerning the lifts at Grenfell Tower', September 2020. https://assets.grenfelltowerinquiry.org.uk/ RHO00000003_Report%20of%20Roger%20Howkins%20 concerning%20the%20lifts%20at%20Grenfell%20Tower%20 %28September%202020%29.pdf

12 Tony Enright, evidence to Australian committee on cladding safety, 6 September 2017. https://www.aph.gov.au/ Parliamentary_Business/Committees/Senate/Economics/ Non-conforming45th/Interim_report_cladding

13 *Phase 1 Report*, volume four, page 597.

14 Michael Dowden, inquiry witness statement.

Chapter 2

1 'Case Study On The Ronan Point Tower Block History Essay'. https://www.ukessays.com/essays/history/case-study-on-the-ronan-point-tower-block-history-essay.php

2 Hilditch, M., 'One man's battle to improve tower block safety', 15 May 2018. https://www.insidehousing.co.uk/insight/insight/ one-mans-battle-to-improve-tower-block-safety-56202

3 Ibid.

4 Ibid.

5 Wearne, P., *Collapse: When Buildings Fall Down* (TV Books Inc, 2000).

6 Hansard, 'Ronan Point and Tower Blocks', 19 February 1985. https://api.parliament.uk/historic-hansard/written-answers/1985/feb/19/ronan-point-and-tower-blocks

7 Charles II, 'An Act for Rebuilding the Citty of London', 1666. https://www.british-history.ac.uk/statutes-realm/vol5/pp603-612

8 FBU, 'The Grenfell Tower Fire: a crime caused by profit and deregulation', 23 September 2019. https://www.fbu.org.uk/publications/grenfell-tower-fire-crime-caused-profit-and-deregulation

9 https://hansard.parliament.uk/Commons/1982-11-23/debates/544b803f-79cf-4c15-88e3-1ce59a78311f/HousingAndBuildingControlBill?highlight=system%20building%20control%20must%20above%20reproach%20building%20control%20officers%20local%20authorities%20independent%20not%20owe%20their%20position%20one%20developer%20that%20must%20continue%20anything%20would%20unacceptable#contribution-65a9cd98-3045-4936-bca1-06c76b75d4f4

10 Lords Hansard, 'Housing and Building Control Bill', 5 April 1984. https://api.parliament.uk/historic-hansard/lords/1984/apr/05/housing-and-building-control-bill-1#division_4

11 Apps, P., 'Special investigation, how the government missed the chance to prevent the cladding crisis', *Inside Housing*, 13 June 2021. https://www.insidehousing.co.uk/insight/insight/special-investigation-how-the-government-missed-the-chance-to-prevent-the-cladding-crisis-in-the-1990s-71109

12 Bisby, L., 'Regulatory testing and the Path to Grenfell', 2022. https://assets.grenfelltowerinquiry.org.uk/LBYP20000001_Professor%20Luke%20Bisby%20Phase%202%20Report%20

-%20Regulatory%20Testing%20and%20the%20Path%20to%20
Grenfell_1.pdf

13   Inquiry transcript, 13 June 2022.

14   Connolly, R. 'Investigation of the behaviour of external cladding
     systems in fire', https://assets.grenfelltowerinquiry.org.uk/
     RCO00000001_Investigation%20of%20the%20behaviour%20
     of%20external%20cladding%20systems%20in%20fire%20-%20
     Report%20on%2010%20full-scale%20fire%20tests%20CR143_
     94%20-%20Fire%20Research%20Station.pdf

15   Morgan, P., Martin, B., Morris, T., 'Fire at Garnock Court,
     Irvine on 11 June 1999', https://assets.grenfelltowerinquiry.
     org.uk/BRE00035377_Report%20by%20the%20BRE%20
     into%20the%20fire%20at%20Garnock%20Court%20in%20
     Irvine%2C%20Scotland%2C%20prepared%20for%20North%20
     Ayrshire%20Council%20by%20Brian%20Martin%20and%20
     others.pdf

16   Memorandum by the Fire Safety Development Group (ROF
     26). https://publications.parliament.uk/pa/cm199899/cmselect/
     cmenvtra/741/9072003.htm

17   Memorandum by Stephen Ledbetter the Centre for Window and
     Cladding Technology (ROF 45). https://publications.parliament.
     uk/pa/cm199899/cmselect/cmenvtra/741/9072011.htm

18   Select Committee on Environment, Transport and Regional
     Affairs First Report, 'Potential risk of fire spread in buildings via
     external cladding systems', https://publications.parliament.uk/
     pa/cm199900/cmselect/cmenvtra/109/10907.htm

19   Brian Martin, 17 March 2022.

20   Interview with author.

21   Brian Martin, inquiry transcript, 17 March 2022: https://
     assets.grenfelltowerinquiry.org.uk/documents/transcript/
     Transcript%2017%20March%202022_0.pdf

22   Kellier, A., 'Letter to Dr Sarah Colwell', 15 August 2002.
     https://assets.grenfelltowerinquiry.org.uk/BRE00042031_
     Letter%20from%20Alan%20Keiller%20%28CWCT%29%20

to%20Sarah%20Colwell%20%28BRE%29%20providing%20
comments%20in%20response%20to%20the%202002%20
consultation%20in%20relation%20to%20the%20revision%20
of%20BR%20135.%20Response%20to%20comments%20are%20
in%20ma.pdf

23 Anthony Burd, inquiry transcript, 28 Feb 2022.

24 Building Regulations Advisory Committee meeting minutes,
April 2002. https://assets.grenfelltowerinquiry.org.uk/
CLG00000720_Minutes%20-%20Building%20Regulations%20
Advisory%20Committee%20Part%20B%20Working%20
Party%20Meeting.pdf

25 RADAR 2 Project, 'Warrington fire research', May 2000.
https://assets.grenfelltowerinquiry.org.uk/CLG00000951_
RADAR%202%20Project%20-%20Research%20Test%20
Report%20-%20Correlation%20of%20UK%20Reaction%20
to%20Fire%2C%20Guidance%20on%20Revision%20of%20
Approved%20Document%20B%20%28Part%202_%20
Proposals%20for%20the%20European%20Supplement%20to%20
Approved.pdf

26 Kingspan, 'New European Fire Classification System, Technical
Bulletin', https://assets.grenfelltowerinquiry.org.uk/KIN000000
60_Kingspan%20document_%20bulletin%20entitled%20
%27New%20European%20Fire%20Classification%20
System%27%20dated%20May%202003.pdf

27 Brian Martin evidence, 22 March 2022.

28 Brian Martin evidence, 22 March 2022.

Chapter 3

1 Purser, D. https://assets.grenfelltowerinquiry.org.uk/documents/
Professor%20David%20Purser%20report%20%28Phase%201%20
-%20supplemental%29%20DAPR0000001.pdf

2 Interview with author.

3 *Phase 1 report*, volume four, page 607.

Chapter 4

1 Inquiry transcript, 10 February 2022.

2 Inquiry transcript, 1 March 2022.

3 Inquiry transcript, 29 March 2022.

4 Cabinet Office press release, 7 April 2011.

5 Barwise, S., 'Opening statement', 6 December 2021. https:// assets.grenfelltowerinquiry.org.uk/BSR00000096_BSR%20 Team%201A%20-%20Phase%202%20Module%206%20 Written%20Opening%20%28Government%2C%20 FRA%2C%20Testing%20and%20Certification%29%20 Submissions%20%5BBindmans%2C%20Hickman%20%26%20 Rose%2C%20Hodge%20Jones%20%26%20Allen%5D.pdf

6 Inquiry transcript, 7 April 2021.

7 Glaze, B., Bentley, D., Woodcock, A., 'David Cameron: I will kill off safety culture', *Independent*, 5 January 2012. https://www. independent.co.uk/news/uk/politics/david-cameron-i-will-kill-off-safety-culture-6285238.html

8 Inquiry transcript, 4 April 2022.

9 Inquiry transcript, 8 March 2022.

10 Inquiry transcript, 9 February 2022.

11 Kirkham, F., 'Letter to DCLG', 28 March 2013. https://www. lambeth.gov.uk/sites/default/files/ec-letter-to-DCLG-pursuant-to-rule43-28March2013.pdf

12 Inquiry transcript, 29 March 2022.

13 Pickles, E., Letter to coroner Frances Kirkham, 20 May 2013. https://www.lambeth.gov.uk/sites/default/files/ec-letter-from-rt-hon-eric-pickles-mp-20May2013.pdf

14 Kirkham F., 'Letter to DCLG', 28 March 2013. https://www. lambeth.gov.uk/sites/default/files/ec-letter-to-DCLG-pursuant-to-rule43-28March2013.pdf

15 Apps, P., 'Grenfell management company ignored Lakanal recommendations after government said they would not become mandatory', *Inside Housing*, 12 June 2019. https://www.

insidehousing.co.uk/news/news/grenfell-management-com-pany-ignored-lakanal-recommendations-after-government-said-they-would-not-be-mandatory-61861

16 Inquiry transcript, 23 March 2022.

17 Letter of Stephen Williams to Sir David Amess, 9 September 2014.

18 Letter of Sir David Amess to Stephen Williams, 28 October 2014.

19 Inquiry transcript, 30 March 2022.

20 Inquiry transcript, 22 March 2022.

21 Inquiry transcript, 24 March 2022.

22 Inquiry transcript, 8 March 2022.

23 Ibid.

24 Inquiry transcript, 28 March 2022.

25 Inquiry transcript, 7 March 2022.

26 Inquiry transcript, 7 April 2022.

27 Inquiry transcript, 10 February 2022.

28 Inquiry transcript, 8 March 2022.

29 Letter of Gavin Barwell to Sir David Amess, 2 May 2017.

30 Inquiry transcript, 25 November 2021.

31 Inquiry transcript, 8 March 2021.

32 Inquiry transcript, 30 March 2022.

33 Ibid.

Chapter 5

1 BSR Team 2 Opening Statements, 30 January 2020. https://www.grenfelltowerinquiry.org.uk/evidence/bsrs-team-2-opening-statements

2 Bisby, L., 'Grenfell Tower Inquiry: Phase 1 – Expert Report', 2 April 2018. https://www.grenfelltowerinquiry.org.uk/evidence/professor-luke-bisbys-expert-report

Chapter 6

1 Court of Appeal of the Supreme Court of Victoria judgement, March 2021.

2 Alcoa, 'A history of innovation', https://www.alcoa.com/global/en/who-we-are/history

3 Inquiry transcript, 16 February 2021.

4 Inquiry transcript, 22 March 2021.

5 Inquiry transcript, 17 February 2021.

6 Ibid.

7 Inquiry transcript, 10 March 2021.

8 Inquiry transcript, 9 November 2020.

9 Inquiry transcript, 10 February 2021.

10 Inquiry transcript, 16 February 2021.

11 Inquiry transcript, 10 February 2021.

12 Inquiry transcript, 14 July 2020.

13 Inquiry transcript, 9 November 2021.

Chapter 7

1 *Phase 1 report*, volume four, page 606.

2 Ibid., page 600.

3 Ibid., page 606.

Chapter 8

1 Tubb, G. and Stylianou, N., 'Long Read: Grenfell Britain's fire safety crisis', *Sky News*, 4 June 2018. https://news.sky.com/story/long-read-grenfell-britains-fire-safety-crisis-11146108

2 Apps, P., 'Hanging in the balance: what is the future for cross-laminated timber?', *Inside Housing*, 29 April 2020.

3 European Phenolic Foam Association, 'Properties of phenolic foam', https://epfa.org.uk/properties-of-phenolic-foam/

4 BRE, 'The Production of Smoke and Droplets from products used to form wall and ceiling linings', 2005. https://www.bre.co.uk/filelibrary/pdf/rpts/partb/ODPM_Smoke_Droplets_Report.pdf

5 Hull, R., et al. 'Fire Behaviour of modern facade materials: understanding the Grenfell Tower fire', *Journal of Hazardous*

Materials, January 2019. https://www.sciencedirect.com/science/article/pii/S0304389418312275?via%3Dihub

6 Institute of Fire Engineers incident directory, 1993 Sun Valley. https://www.ife.org.uk/Firefighter-Safety-Incidents/sun-valley-1993/34014

7 Apps, P., 'Special investigation: how the government missed the chance to prevent the cladding crisis in the 1990s', *Inside Housing*, 13 June 2021. https://www.insidehousing.co.uk/insight/insight/special-investigation-how-the-government-missed-the-chance-to-prevent-the-cladding-crisis-in-the-1990s-71109

8 *Fire Magazine*, 1997.

9 Memorandum by the Fire Brigades Union (ROF 28). https://publications.parliament.uk/pa/cm199899/cmselect/cmenvtra/741/9072002.htm

10 Knowsley Heights file, accessed at the National Archives. From letters between John Southern and Tony Morris in 1992.

11 Environment, Transport and Regional Affairs select committee, minutes of evidence, 20 July 1999, questions 40-59. https://publications.parliament.uk/pa/cm199899/cmselect/cmenvtra/741/9072008.htm

12 Celotex financial statements 2005/6.

13 Kingspan financial statements 2005/6.

14 Inquiry transcript, 23 November 2020.

15 Kingspan, 'What's lurking behind your facade flyer', May 2005.

16 Inquiry transcript, 30 November 2020.

17 'Celotex pink bus completes "insulating Britain" promotional tour', *East Anglian Daily Times*, 21 June 2014. https://www.eadt.co.uk/news/business/hadleigh-celotex-pink-bus-completes-insulating-britain-promotional-tour-2146138

18 Inquiry transcript, 19 November 2020.

19 Inquiry transcript, 21 September 2020.

20 Inquiry transcript 23 March 2022.

21 Ibid.

22 Ibid.

23 Inquiry transcript 3 December 2020.

24 Inquiry transcript, 4 December 2020.

25 Inquiry transcript, 14 December 2021.

26 Inquiry transcript, 24 November 2020.

27 NHBC, 'Acceptability of common wall constructions containing combustible materials in high-rise buildings', July 2016. https://assets.grenfelltowerinquiry.org.uk/KIN00000516_Exhibit%20AP_4%20-%20Adrian%20Pargeter%20%28Kingspan%29.pdf

28 Inquiry transcript, 21 September 2020.

29 Inquiry transcript, 9 June 2022.

30 Inquiry transcript, 9 November 2020.

31 Inquiry transcript, 20 June 2022.

Chapter 9

1 *Phase 1 report*, volume four, page 619.

Chapter 10

1 Interview with the author, August 2020.

2 Boughton, J., 'A perfect storm of disadvantage: the history of Grenfell Tower', *The i*, 26 July 2017. https://inews.co.uk/news/perfect-storm-disadvantage-history-grenfell-tower-80807

3 Interview with author, August 2020.

4 Notting Barns South Draft Masterplan Report, July 2009. https://grenfellactiongroup.files.wordpress.com/2015/08/notting-barns-south-masterplan.pdf

5 Eddie Daffarn witness statement, *Phase 2 report*.

6 Inquiry transcript, 3 March 2020.

7 Inquiry transcript, 3 March 2020.

8 Apps, P., 'Grenfell cladding consultation did not mention fire safety', *Inside Housing*, 11 July 2017. https://www.insidehousing.co.uk/news/news/grenfell-cladding-consultation-did-not-mention-fire-safety-51419

9 Booth, R., 'Grenfell Tower: fire-resistant cladding plan was dropped', *Guardian*, 8 May 2018. https://www.theguardian.

com/uk-news/2018/may/08/grenfell-tower-more-costly-fire-resistant-cladding-plan-was-dropped

10 Interview with author.

11 Exova, 'Grenfell Tower Outline Fire Safety Strategy', 31 October 2012. https://assets.grenfelltowerinquiry.org.uk/EXO00000519_Exova%20Grenfell%20Tower%20Outline%20Fire%20Safety%20Strategy%20dated%2031_10_2011.pdf

12 Inquiry transcript, 9 July 2020.

13 Inquiry transcript, 7 October 2020.

14 Ibid.

15 Inquiry transcript, 9 September 2020.

16 Inquiry transcript, 28 July 2020.

17 Inquiry transcript, 15 October 2020.

18 Inquiry transcript, 7 September 2020.

19 Inquiry transcript, 9 July 2020.

20 Apps, P., 'Hundreds of building control surveyor posts cut by councils since 2010, research reveals', *Inside Housing*, 28 April 2021. https://www.insidehousing.co.uk/news/news/hundreds-of-building-control-surveyor-posts-cut-by-councils-since-2010-research-reveals-70525

21 Inquiry transcript, 10 March 2021.

22 Inquiry transcript, 1 October 2020.

23 Inquiry transcript, 30 September 2020.

24 Inquiry transcript, 16 September 2020.

25 Inquiry transcript, 9 March 2021.

26 Inquiry transcript, 17 September 2020.

27 Inquiry transcript, 29 September 2020.

28 Channel 4, *Grenfell: The Untold Story*, broadcast on 8 September 2021.

29 Inquiry transcript, 18 May 2021.

30 Transcript of Eddie Daffarn speech, provided to author.

31 Inquiry transcript, 19 May 2021.

32 Ibid.

33 Royal Borough of Kensington and Chelsea, 'North Kensington Tower Block transformed by £10m. Refurbishment', 13 May 2016. https://www.rbkc.gov.uk/press-release/ north-kensington-tower-block-transformed-£tom-refurbishment

34 Inquiry transcript, 1 October 2020.

Chapter 11

1 Generic Risk Assessment 3.2 and LFB Policy No. 633 par 7.46. https://www.highrisefirefighting.co.uk/docs/GRA%203.2%20 High%20rise.pdf

2 Purser D., 'Phase 1 Report: General description of hazards excluding comprehensive references to individual occupants', 5 November 2018. https://www.grenfelltowerinquiry.org.uk/ evidence/professor-david-pursers-expert-report

Chapter 12

1 Council papers accessed from Kensington library.

2 Memoli, M., 'Investigation report on longstanding complaints of the Kensington and Chelsea TMO', April 2009. https://assets. grenfelltowerinquiry.org.uk/IWS00001462_Exhibit%20 SA_8%20-%20Investigation%20report%20on%20the%20long- standing%20complaints%20of%20the%20KCTMO%20by%20 Maria%20Memoli.pdf

3 My London, 'New Kensington and Chelsea TMO boss promises radical shake-up', 1 June 2009. https://www.mylondon.news/ news/local-news/new-kensington-chelsea-tmo-boss-6007382

4 Interview with author.

5 Barker, N., 'KCTMO left thousands of repairs undone', *Inside Housing*, 12 January 2018. https://www.insidehousing.co.uk/ news/news/kctmo-left-thousands-of-repairs-undone-council- papers-reveal-53919

6 Inquiry transcript, 20 April 2021.

7 Shah Ahmed, witness statement, *Phase 2 report*.

8 Inquiry transcript, 27 April 2020.

9 Inquiry transcript, 21 April 2020.

10 Interview with the author.

11 Barnes, S., 'The biggest ever survey of fire risk assessments has revealed widespread safety problems', *Inside Housing*, 13 June 2018. https://www.insidehousing.co.uk/insight/insight/the-biggest-ever-survey-of-fire-risk-assessments-has-revealed-widespread-safety-problems-56774

12 Booth, R., 'Grenfell Tower door resisted fire for half as long as it was meant to', *Guardian*, 15 March 2018. https://www.theguardian.com/uk-news/2018/mar/15/grenfell-tower-door-resisted-fire-half-as-long-as-it-was-meant-to

13 Inquiry transcript, 20 April 2021.

14 Natasha Elcock, Phase 1 witness statement.

15 Inquiry transcript, 19 April 2021.

16 Seamus Dunlea, witness statement.

17 Inquiry transcript, 27 May 2021.

18 Harley, N., '50 Rescued from burning flats in Kensington', *Telegraph*, 31 October 2015.

19 Inquiry transcript, 29 April 2021.

20 Inquiry transcript, 12 May 2021.

21 Inquiry transcript, 10 June 2021.

22 Apps, P. and Barker, N., 'The Grenfell Tower Inquiry report: one year on, is the sector acting on the recommendations?', *Inside Housing*, 30 October 2020. https://www.insidehousing.co.uk/insight/insight/the-grenfell-tower-inquiry-report-one-year-on-is-the-sector-acting-on-the-recommendations-68316

23 Apps, P., 'Moratorium on sales of composite doors lifted', *Inside Housing*, 11 December 2018. https://www.insidehousing.co.uk/news/news/moratorium-on-sale-of-composite-fire-doors-lifted-59467.

24 Barker, N., 'Three-quarters of composite fire doors failed safety tests', *Inside Housing*, 14 February 2019. https://www.insidehousing.co.uk/news/three-quarters-of-composite-fire-doors-failed-safety-tests-60190

25 Apps, P., 'Fire doors: a systemic problem?', *Inside Housing*, 26
    July 2018. https://www.insidehousing.co.uk/insight/insight/
    fire-doors-a-systemic-problem-57326

26 Documents provided to author.

27 Hewitt, D., 'Britain's Housing Shame: a story of shocking con-
    ditions and tenants' despair at lack of action', *ITV News*, 12
    September 2021. https://www.itv.com/news/2021-09-12/brit-
    ains-housing-shame-shocking-conditions-and-despair-at-a-lack-
    of-action

28 Brown, C., 'Shapps confirms plan to scrap TSA', *Inside Housing*,
    24 June 2010. https://www.insidehousing.co.uk/news/news/
    shapps-confirms-plans-to-scrap-tsa-20665

29 Hardman, I., 'Death of a watchdog', *Inside Housing*, 25 June 2010.
    https://www.insidehousing.co.uk/insight/insight/death-of-
    a-watchdog-20572

30 Hansard, 'Circle Housing and Orchard Village', 12 January 2017.
    https://hansard.parliament.uk/Commons/2017-01-12/debates/
    17011264000002/CircleHousingAndOrchardVillage

31 Leslie Thomas QC, submission to Grenfell Tower Inquiry, 7 July
    2021.

32 Snaith, E., 'Council workers called Grenfell area "little
    Africa" after deadly fire, MP says', *Independent*, 7 June 2019.
    https://www.independent.co.uk/news/uk/home-news/
    grenfell-kensington-chelsea-council-little-africa-tropics-emma-
    dent-coad-a8948526.html

33 Dent Coad, E. 'The most unequal borough in Britain – revisited',
    October 2020. https://www.dropbox.com/s/87fjaogenl945ws/
    The%20Most%20Unequal%20Borough%20in%20Britain%20
    21.10.20.pdf?dl=0

34 Eddie Daffarn, *Phase 2 report* witness statement.

Chapter 13

1 Interview with author, January 2021.

Chapter 14

1 Matt Wrack *Phase 2 report* witness statement.
2 Todd, C., 'Legislation, guidance and enforcing authorities relevant to fire safety measures at Grenfell Tower', March 2018. https://assets.grenfelltowerinquiry.org.uk/documents/Colin%20 Todd%20report_0.pdf
3 Inquiry transcript, 14 March 2022.
4 Apps, P., 'The Secret Life of a Risk Assessor: whistleblower warns of culture of cover-up', *Inside Housing*, 21 March 2022. https://www.insidehousing.co.uk/insight/insight/ the-secret-life-of-a-fire-risk-assessor-whistleblower-warns-of-culture-of-cover-up-74400
5 Barnes, S., 'Revealed: the most common fire safety problems in tower blocks', *Inside Housing*, 4 August 2017. https://www. insidehousing.co.uk/insight/insight/revealed-the-most-common-fire-safety-problems-in-tower-blocks-51730
6 Janice Wray evidence, 7 June 2021.
7 Barratt, L., 'KCTMO appointed "competitively priced" fire risk assessment consultant', *Inside Housing*, 16 June 2018. https:// www.insidehousing.co.uk/news/news/the TMO-appointed-competitively-priced-fire-risk-assessment-consultant-50993
8 Inquiry transcript, May 26 2021.
9 Inquiry transcript, April 28 2021.
10 Inquiry transcript, 27 May 2021.
11 Ibid.

Chapter 15

1 Inquiry transcript, 15 November 2018.
2 Ibid.

Chapter 16

1 Todd, C. 'Legislation, guidance and enforcing authorities relevant to fire safety measures at Grenfell Tower', March 2018. https://

assets.grenfelltowerinquiry.org.uk/documents/Colin%20
Todd%20report_0.pdf

2 Holland C., et al., 'External Fire Spread – Part 1 Background
   research', April 2016. https://www.bre.co.uk/filelibrary/Fire%20
   and%20Security/FI---External-Fire-Spread-Part-1.pdf

3 Apps, P., 'Where did the stay put policy come from and where
   do we go now?', *Inside Housing*, 31 October 2019. https://www.
   insidehousing.co.uk/insight/insight/where-did-the-stay-put-
   policy-come-from-and-where-do-we-go-now-63957

4 Hodinkson, S., Murphy, P., Turner, A., 'The Fire Risks of
   Purpose Built Blocks: an exploration of incident data in
   England', July 2021. https://d3bbcf31-25c6-489a-ad4f-
   3e78dca61563.usrfiles.com/ugd/d3bbcf_637ac7cb24b547828ff1
   952056dd60a3.pdf

5 Inquiry transcript, 21 October 2021.

6 Inquiry transcript, 20 September 2021.

7 C.S. Todd & Associates, 'Fire Safety in Purpose Built Blocks of
   Flats', 2011. https://assets.publishing.service.gov.uk/govern-
   ment/uploads/system/uploads/attachment_data/file/1020410/
   Fire_Safety_in_Purpose_Built_Blocks_of_Flats_Guide.pdf

8 Ibid.

9 Inquiry transcript, 27 July 2021.

10 Inquiry transcript, 15 March 2022.

11 Ibid.

12 Inquiry transcript, 28 July 2021.

13 Apps, P., 'How social landlords are failing to provide evacuation
    plans for disabled residents', *Inside Housing*, 27 September 2021.
    https://www.insidehousing.co.uk/insight/insight/how-social-
    landlords-are-failing-to-prepare-emergency-evacuation-plans-
    for-disabled-residents-72430

14 Inquiry transcript, 23 June 2021.

15 Inquiry transcript, 29 March 2021.

16 Kirkham, F., 'Rule 43 letter to DCLG', 28 March 2013. https:// www.lambeth.gov.uk/sites/default/files/ec-letter-to-DCLG- pursuant-to-rule43-28March2013.pdf

17 Pickles, E., 'Letter to Frances Kirkham', 20 May 2013.https:// www.lambeth.gov.uk/sites/default/files/ec-letter-from-rt- hon-eric-pickles-mp-20May2013.pdf

18 Inquiry transcript, 3 November 2021.

19 Ibid.

20 Apps, P., 'Grenfell five years on: could it happen again?', *Inside Housing*, 13 June 2021. https://www.insidehousing.co.uk/ insight/grenfell-five-years-on-could-it-happen-again-76000

21 Home fire safety guide for purpose-built flats and maisonettes, London Fire Brigade, distributed at Plaistow Fire Station on October 31 2021.

22 Inquiry transcript, 15 June 2022.

23 Apps, P., 'Grenfell five years on: could it happen again?', *Inside Housing*, 13 June 2021. https://www.insidehousing.co.uk/ insight/grenfell-five-years-on-could-it-happen-again-76000

Chapter 18

1 Narrative verdict into death of Catherine Hickman, 28 March 2013. https://www.lambeth.gov.uk/sites/default/files/ec-inqui- sition-and-narrative-verdict-catherine-hickman.pdf

2 Kirkham, F., 'Letter to LFB pursuant to Rule 43', 28 March 2013. https://www.lambeth.gov.uk/sites/default/files/ec-letter- to-london-fire-brigade-pursuant-to-rule43-28March2013.pdf

3 Inquiry transcript, 16 November 2021.

4 Inquiry transcript, 27 September 2018.

5 Inquiry transcript, 5 October 2021.

6 Inquiry transcript, 20 September 2021.

7 Baigent, D., 'One more last working class hero: a cultural audit of the UK Fire Service', 2001. https://www.academia.edu/4254126/

ONE_MORE_LAST_WORKING_CLASS_HERO_A_
CULTURAL_AUDIT_OF_THE_UK_FIRE_SERVICE

8 Inquiry transcript, 20 September 2021.
9 Knight, K., 'Facing the future', May 2013. https://www.gov.uk/
   government/publications/facing-the-future
10 National Audit Office, 'Impact of funding reductions on fire and rescue services', November
   2018. https://www.nao.org.uk/report/
   impact-of-funding-reductions-on-fire-and-rescue-services/
11 Hansard, 30 October 2019. https://hansard.parliament.uk/com
   mons/2019-10-30/debates/97A7B2AB-DD4E-427D-
   BBBF-431A1B8E7017/GrenfellTowerInquiry
12 BBC News, 'Mayor Boris Johnson tells opponent to get
   stuffed', 11 September 2013. https://www.bbc.co.uk/news/av/
   uk-england-london-24050870

Chapter 19

1 Inquiry transcript, 13 April 2022.
2 Inquiry transcript, 25 April 2022.
3 Inquiry transcript, 13 April 2022.
4 Ibid.
5 Inquiry transcript, 26 April 2022.
6 Inquiry transcript, 10 April 2022.
7 Inquiry transcript, 10 April 2022.
8 Interview with author.

Chapter 20

1 Mortimer, C., 'Camden residents face third night in leisure
   centre amid Grenfell cladding aftermath', Independent, 27 June
   2017. https://www.independent.co.uk/news/uk/home-news/
   chalcots-estate-camden-council-cladding-fire-risk-grenfell-
   tower-georgia-gould-a7807801.html
2 Stewart, H., 'Theresa May announces public inquiry into Grenfell
   Tower fire', Guardian, 15 June 2017. https://www.theguardian.

com/uk-news/2017/jun/15/theresa-may-announces-public-inquiry-into-grenfell-tower-fire

3 Inquiry transcript, 16 March 2022.

4 Inquiry transcript, 23 March 2022.

5 Inquiry transcript, 3 February 2022.

6 Inquiry transcript, 30 March 2022.

7 Orr, D., 'Ministers need to take new approach in Grenfell response', *Inside Housing*, 30 June 2017. https://www.insidehousing.co.uk/comment/comment/ministers-need-to-take-a-new-approach-in-grenfell-response-51259

8 Ministry of Housing, Communities & Local Government, 'Expert panel appointed to advise on immediate safety action following Grenfell fire', 27 June 2017. https://www.gov.uk/government/news/expert-panel-appointed-to-advise-on-imme-diate-safety-action-following-grenfell-fire

9 Advice Note 11, expert advisory panel: https://assets.grenfell-towerinquiry.org.uk/documents/60.%20Building%20Safety%20Programme%20update%20and%20consolidated%20advice%20for%20building%20owners%20following%20large-scale%20testing%20-%20exhibit%20to%20MHCLG_CLG10003157.pdf

10 Interview with author.

11 Simpson, J., 'Birmingham high rise with fire safety issues faces 1,237% insurance premium hike', *Inside Housing*, 27 April 2020, https://www.insidehousing.co.uk/news/news/birmingham-high-rise-with-fire-safety-issues-faces-1237-insurance-premium-hike-66215

12 Barnes, S., 'Fire safety: the leaseholder issue', *Inside Housing*, 2 March 2018. https://www.insidehousing.co.uk/insight/insight/fire-safety-the-leaseholder-issue-54918

13 De Gallier, T., 'Imprisoned by cladding: The flat owners who cannot sell', 8 February 2020. https://www.bbc.co.uk/news/stories-51412328

14 Apps, P., 'Revealed: the mental health trauma of residents in private blocks with dangerous cladding', *Inside Housing*,

26 April 2019. https://www.insidehousing.co.uk/insight/
revealed-the-mental-health-trauma-of-residents-in-private-
blocks-with-dangerous-claddingi-61169

15 Simpson, J., 'Barking fire: the inside story', *Inside Housing*, 13
September 2019. https://www.insidehousing.co.uk/insight/
insight/barking-fire-the-inside-story-63110

16 Apps, P., 'Swift evacuation of the Cube "saved many lives,"
says fire report', *Inside Housing*, 31 July 2020. https://www.
insidehousing.co.uk/news/news/swift-evacuation-of-the-cube-
saved-many-lives-says-fire-report-67357

17 Siddle, J., 'First suicide victim linked to cladding scandal
feared huge bills and no way out', *Daily Mirror*, 3 October
2021. https://www.mirror.co.uk/news/uk-news/
first-suicide-victim-linked-cladding-25124910

18 Apps, P., 'Grenfell five years on: could it happen again?', *Inside
Housing*, 13 June 2021. https://www.insidehousing.co.uk/
insight/grenfell-five-years-on-could-it-happen-again-76000

19 Inquiry transcript, 25 March 2021.

20 Barratt, L. and Apps, P., 'What do the leaked Kingspan minutes
show?', *Inside Housing*, 22 February 2018. https://www.
insidehousing.co.uk/insight/insight/what-do-the-leaked-
kingspan-meeting-notes-show-54739

21 Barratt, L., 'The Hackitt Review: key recommendations at a
glance', *Inside Housing*, 17 May 2018. https://www.insidehousing
.co.uk/insight/insight/the-hackitt-review-key-
recommendations-at-a-glance-56337

22 Apps, P., 'Grenfell survivors "saddened and disappointed" by
Hackitt report', *Inside* Housing, 17 May 2018. https://www.
insidehousing.co.uk/news/news/grenfell-survivors-saddened-
and-disappointed-by-hackitt-report-56329

23 Apps, P., 'Three quarters of cladding systems on new medium
rise buildings use combustible materials', *Inside Housing*, 6
April 2021. https://www.insidehousing.co.uk/news/news/

three-quarters-of-cladding-systems-on-new-medium-rise-buil-dings-use-combustible-materials-data-shows-70298

24 *Phase 1 report*, volume four.

25 Apps, P., 'Government watered down implementation of Grenfell recommendations for disabled people after push from lobbyists', *Inside Housing*, 18 December 2020. https://www.insidehousing .co.uk/news/news/government-limited-grenfell-inquiry-rec-ommendations-for-disabled-people-after-push-from-industry-lobbyists-69036

26 Apps, P., 'Government's rejection of proposals for disabled resi-dents branded "shameful" and "reprehensible"', *Inside Housing*, 19 May 2022. https://www.insidehousing.co.uk/news/news/ governments-rejection-of-grenfell-inquiry-proposals-for-disa-bled-residents-branded-shameful-and-reprehensible-75683

27 Apps, P., 'Are two fires in the Shetland Islands a canary in the coal mine for modular housing?', *Inside Housing*, 16 October 2020. https://www.insidehousing.co.uk/insight/insight/are-two-fires-on-the-shetland-islands-a-canary-in-the-coal-mine-for-modular-construction-68170

Chapter 21

1 Lomas, C., 'Hundreds of children struggling with mental health issues after fire', *Sky News*, 11 June 2018. https://news.sky.com/ story/hundreds-of-children-struggling-with-mental-health-issues-after-grenfell-11401162

2 Bowden, G., 'Toxic Grenfell cough leaves survivors and fire-fighters with health problems, MPs say', *Huffington Post*, 16 July 2019. https://www.huffingtonpost.co.uk/entry/grenfell-cough_ uk_5d2cbf9ce4b08938b09922ea#:~:text=An%20emerging%20 %E2%80%9CGrenfell%20cough%E2%80%9D%20has,Audit%20 Committee%20(EAC)%20said.

3 Barratt, L., 'Grenfell's forgotten victims: life on the Lancaster West estate', *Inside Housing*, 14 June 2019. https://www.insidehousing.

co.uk/insight/insight/grenfells-forgotten-victims-life-on-the-lancaster-west-estate-after-the-fire-61817

4 Barratt, L., 'Block of flats chosen to house Grenfell survivors has high fire risk', *Inside Housing*, 14 August 2018. https://www.insidehousing.co.uk/news/news/block-of-flats-chosen-to-house-grenfell-survivors-found-to-have-high-fire-risk-62729

5 Independent Grenfell Recovery Taskforce Third Report, 21 November 2018. https://assets.publishing.service.gov.uk/government/uploads/system/uploads/attachment_data/file/949654/Grenfell_Recovery_Taskforce_Third_Report.pdf

6 Hosken, A., 'Inspectors find "appalling" fire risks at east London tower block', *BBC News*, 23 June 2017. https://www.bbc.co.uk/news/av/uk-40382636

7 Simpson, J., 'Around 150 people forced to evacuate after fire at block which recently had ACM cladding removed', *Inside Housing*, 23 June 2020. https://www.insidehousing.co.uk/news/news/around-150-people-forced-to-evacuate-after-fire-at-block-which-recently-had-acm-cladding-removed-66932

Chapter 22

1 *Phase 1 report*, page 729.

Peter Apps is an award-winning journalist and Deputy Editor at *Inside Housing*. He broke a story on the dangers of combustible cladding thirty-four days before the Grenfell Fire. He has not stopped reporting on this national tragedy since, and his coverage of the public inquiry has received widespread acclaim. He lives in London.

WHAT YOU CAN DO

There are many groups fighting for justice for the Grenfell bereaved and survivors and broader change. Grenfell United is a large collective of bereaved and survivors; Justice for Grenfell is a community-led justice campaign and Grenfell Next of Kin represents some of the bereaved families. The End Our Cladding Scandal campaign is a national collective of residents who live in blocks with dangerous cladding and other safety defects. All of these groups organise protests and social media campaigns and can be found online. Claddag (Cladding Disability Action Group) campaigns specifically on behalf of disabled residents in high-rise buildings. They have been supported by groups including Disability Rights UK and continue to advocate for evacuation plans for residents with disabilities. Tower Blocks UK is a high-rise safety campaign founded by activists including Sam Webb and aims to support social housing residents in dangerous blocks. The London Renters Union and ACORN network support the rights of private and social housing tenants. The North Kensington Law Centre, Al-Manaar Mosque, Clement St James Centre and Rugby Portobello Trust are among the organisations continuing to support communities in the area around the tower. A proportion of the author's proceeds from this book will be used to support them and other grassroots local charities.